ALSO BY LEW W. L. SALLEDYN

Fragments of the History of Assyria

Scientific Fragments
Songs of Mourning

Lew W. L. Salledyn
salledyn@gmail.com

Scapegrace and Misfit Press 2017

Note to a spectator

This is a work neither of scholarship nor of art – or it is a work almost of scholarship, almost of art – written by a person originally trained in the natural sciences who, alarmed by what he saw there, spent a decade reading and thinking and only occasionally writing. Were this a work of scholarship, it would display a mastery of the recent academic literature, particularly in what are called the humanities, which it does not. Were this a work of art, it would disdain what fragile and transitory scholarship it deploys. As a work of hybridity, however, it endeavors to exemplify, even as failed prototype, what must come; but it is, more than a work of hybridity, a work of almost – and in that it is necessarily not only hybrid but monstrous; for necessarily, a work of almost will always have been premature.

Contents

Prologue

"...[O]nly Agathon, Aristophanes, and Socrates[1] *were still awake, drinking out of a large cup which they were passing around from left to right. Socrates was talking to them. Aristodemus couldn't remember exactly what they were saying – he'd missed the first part of their discussion, and he was half-asleep anyway – but the main point was that Socrates was trying to prove to them that authors should be able to write both comedy and tragedy: the skillful tragic dramatist should also be a comic poet. He was about to clinch his argument, though, to tell the truth, sleepy as they were, they were hardly able to follow his reasoning. In fact, Aristophanes fell asleep in the middle of the discussion, and very soon thereafter, as day was breaking, Agathon then drifted off. But after getting them off to sleep, Socrates got up and left, and Aristodemus followed him, as always. He said that Socrates went directly to the Lyceum, washed up, spent the rest of the day just as he always did, and only then, as evening was falling, went home to rest."*

Plato, the concluding lines of the Symposium

A friend of mine recently suggested that the hermeneutic environment in which we find ourselves is calling for the articulation of a "tragic science." Further, he clarified, this would not be nihilism but would oppose it; "tragedy" would be understood here as an ontological paradigm with roots in

---

[1] The historical figures Agathon and Aristophanes were respectively a tragic and comic poet. Throughout the *Symposium* Socrates is associated with satyrs, both mythological figures and a third type of ancient Greek drama described below.

the tragic poets of ancient Greece. It might seem, from my title, that I agree with him. What are songs of mourning if not tragic? Indeed, the reader of this book will encounter tragedy in one of its traditional senses – as culminating, always, in death – as well as in one of its most literal senses – as a goat (τράγος, tragos) song (αοιδή, aoidē). For more than one song of a dead or dying goat will be found in these pages, and it is not a trivial concern of this book, even if it is unavoidably an unresolved one, whether the song belongs to the goat or the goat to the song: that is, whether the song of a goat is a subjective or objective genitive. This enduring ambiguity between subject and object, self and other, human and non-human, is an aspect of these songs' mournfulness, and the question, "Have I heard a goat sing?" may be among this book's most serious. And if I have, has it been at the expense of other songs – the songs of rats and ants, of tortoises and rails, of dandelions and *Scalesia*, for example?

Am I kidding? Can I seriously expect a reader to bend forward, listen intently and wait, perhaps forever, for a goat to sing, not to mention a rat, a dandelion or a fat lady in a cab?[2] Perhaps what I present here is not tragic at all, but comic: perhaps these are songs of revelry, songs of morning with which villagers greet the dawn. Are we on the cusp of a day, a wedding day even, when the sciences and the humanities call

---

2 "But everyone remembered; what she loved was this, here, now, in front of her; the fat lady in the cab. Did it matter then, she asked herself, walking towards Bond Street, did it matter that she must inevitably cease completely; all this must go on without her; did she resent it; or did it not become consoling to believe that death ended absolutely?" Virginia Woolf, Mrs. Dalloway, p. 9.

an end to their overdetermined estrangement, their flirtatious bickering, and yield to their mutual passion? What, if anything, will be born of that union, we cannot say; whether it will be a sterile hybrid, or a monster, or an abortion, or simply a child, we cannot say; whether a god or a beast – or something else – we cannot say. But if we would get past the sciences, past the humanities, if we would come to what comes after, it will only be by what they beget together. A post-humanities, absent a post-sciences, is a twin only half born, only half a twin: and half a twin is no more a twin than half a circle is a circle. It is comic, it is laughable, to announce the progeny in advance of the begetting. Much of what I recount here is nothing but that comedy of errors, that hilarious and self-criticizing projection of desire which is legible for now only to the audience and an occasional slave, and whose culmination, tradition promises, will eventually be a marriage. Can tragic science, then, also be comic? And if it is truly comic, will it still be science, or will it be what comes after?

The satyr, for example – that hybrid of goat and man and god, moreover to which Socrates in the *Symposium* is constantly compared; and the final play, as it were the last act, following the trilogy of tragedies performed by Athenians in honor of Dionysus, an unusually mortal god. The satyr dramas were supposedly tragicomic, bawdy, burlesque, often entailing a laughable suffering, a small, orgasmic death. So we are told; but the testimony is fragmentary. Only one satyr drama, the *Cyclops* of Euripides, has survived more or less intact. It is the story of that consummate wanderer Odysseus, he who goaded men to follow knowledge to the ends of the earth,[3] and who died, Dante tells us, within sight of a heaven-ascending mountain which undoubtedly he would have endeavored to climb. But the satyr drama which survived, more or less, does not tell us this story; it tells us instead about how the Greek hero blinded the Cyclops. In Homer, he escaped because the blinded Cyclops could not distinguish him from an animal; but Euripides again tells a different story, a story always subtly different from what we expect. His Odysseus could have escaped without violence. His Odysseus chose not to. The blinding of the Cyclops is unnecessary, gratuitous, one might even say funny if it's appropriate to laugh at such things. In its superfluity, we suspect, it is either meaningless or excessively meaningful. Here cleverness (μητις) either falls short or exceeds itself. But which? Odysseus does not resolve this question for us; he does not explain himself. He merely declares himself author of the act whose meaning, if it has any, we do not know. Nor, perhaps, does he.

Why has this been done? "That," Odysseus answers, "is a metaphysical question – beyond our purview here. But I can answer a different question: who did it?" He pauses expectantly, dramatically. "Μητις" – cleverness, no one – "did it." Again he pauses, now smiling or even leering. "But I, Odysseus, am cleverness (μητις). I, Odysseus, am no one (Μητις)!"[4]

---

[3] In this he was precursor to Francis Bacon, whose title page to The Great Instauration portrayed a ship sailing past the Pillars of Hercules.

[4] Μητις and μητις would both be transcribed as "Mētis" in the English alphabet, with the difference being only whether the initial m (mu) is capitalized.

Was Odysseus deaf, was he blind (or blinded), to the ambiguities of meaning in his own confession? Did he sense how this confession or declaration repeated the gratuitous, superfluous act whose authorship it claimed? Would meaning exceed itself, be hyperbolic; would it fall short, be elliptical; or does meaning's parabolic trajectory find its endless completion in the parable? Can no one be so clever as to author?

Perhaps the work of this book is to hear Odysseus' claim in three dimensions at once: the tragic (which, according to Aristotle, represents people as better than in actual life), the comic (which, Aristotle says, represents people as worse than in actual life) and the satyric (but what does Aristotle say about satyrs?). Perhaps as well we should listen for it in unexpected places. For example – from the lips of a goat. Should we ever hear the song of a goat, and should we ever begin to understand it, might the goat not sing, "I too am Μητις"? And we, astonished, will cry and, delighted, will laugh and then – what then? Will we begin to see how we could so often confuse ourselves with goats, how we could mingle with them, destroy them, love them and hate them; how divine figures and devilish ones could be born of our union with them?

And yet the next step, the step beyond, would it not also insist that, even if we are all Μητις – you and me and the goats, and perhaps others as well – even if everyone is no one, still we are not all the same? For, cleverness may be cleverness, but no one is no one. Not even no one is no one.

Parodos (πάροδος): On Superheroes and Alter Egos, Science and Humanism

αἴλινον αἴλινον εἰπέ[1]

In what Robert Burton described as a bastard age of banished muses, the Wittenberg scholar Hamlet toiled to know the world through fiction, only in the last act to yield himself to a story and his body to the stage. Hamlet's trials in benighted Denmark exemplify the angst typically found at the core of the liberal arts and especially the humanities. Indeed, humanists at their best are Hamlet's heirs and, like him, are constantly waylaid by the uncertain responsibilities of a no-less-certain inheritance. Traditionally, they also misidentify and misplace their angst. A modern form of that parapraxis is the fashionable lamentation over the "crisis in the humanities." To be clear, I do not deny that the humanities are in crisis. University departments in the humanities are being underfunded, shuttered, diverted to seemingly pragmatic purposes or degraded to ideological echo chambers. Much of this is driven by irritatingly stupid social, economic and political assessments of the lack of value of a humanistic education. Would-be defenders have, in turn, not very intelligently girded their loins to insist on that value – often conceding the argument in their acceptance of its terms. Lost in the ado is a recognition that the humanities have always been in crisis, will always be in crisis, and are defined by crisis

1 So repeatedly sings the chorus in Aeschylus' *Agamemnon*, for example at line 121. "Sing sorrow, sorrow..."

– beginning with Socrates' deferral of what might be called natural scientific pursuits in order to examine his own potential monstrosity (*Phaedrus* 229c-230a), climaxing in Hamlet's attempt to welcome what is stranger in heaven and earth than dreamt of by philosophy (Act I, scene 5), and only mistakenly identified with the poor worm wriggling in the dirt of modern American academia far beneath the banner "Humanities in Crisis."[2] Like Socrates, we do not know what we are.[3] In this light, the fate of humanities departments is at best our perplexity's aborted form. Its major accomplishment is to afford an illusion, however pessimistic, of possible passage where properly there is none.

Strikingly, the natural sciences are not generally understood to be in crisis, the assessments of Weber, Husserl and Heidegger notwithstanding. Rather, the sciences are the successful younger sibling to which the misfit humanities are routinely and unfavorably compared. This reflects the sciences' perceived productivity, utility, alliances with technology, industry and capitalism, and reduction of knowledge to information – i.e., their overall if unexamined "progress." The sciences, not beset with the ruminative self-doubt and insecurity which characterize at its best the honesty of humanistic inquiry, would instead be the cure to what, from their vantage, looks like the humanities' hypochondriacal

---

[2] I recall Mallarmé's alarm, expressed in an 1894 Oxford lecture, at an earlier crisis in the humanities: "I am indeed bringing you news... Nothing like it ever happened before. *On a touché au vers* (They have tampered with [the rules of] verse/the worm)."

[3] We do not know what face (πρόσωπον) lies before our eyes (προσ ωποῖν) beneath the mask (προσωπεῖον) – or whether a face lies there.

predilection for crises. Such care is palliative. It often culminates in efforts that recast the humanities in science's image, either as a handmaid to narrate science's triumphalist account – to which many philosophies and histories of science reduce – or as a "consilience" whereby a synthesis of science and the humanities is achieved through the latter's gross impoverishment (or a grossly impoverished understanding of the latter). My hope, in contrast, is not to alleviate the crisis in the humanities but to exacerbate and mobilize it; my goal is to infect modern science, particularly biology, with the aporia out of which the best humanistic traditions are born and to which they persistently return.[4]

Aporetic infection has always been a retrospective diagnosis, however. To infect the sciences would be to reveal how the sciences are already infected – how latent within them like a virus is their own critique and crisis. For a new, eventually post-humanistic science to emerge, that critique must be read and that crisis realized. Such work requires a willingness to exemplify and to forge an interdisciplinarity which for the most part has yet to exist, in which the humanities and sciences are not opposed to, but rather are self-consciously included within, one another. The project, then, is not to translate – that is, to carry across, make a passage – the sciences into a humanistic idiom, but instead to bring the resources of the humanities toward the sciences. Precisely what is it that the sciences are always carrying but not carrying

[4] In using the word "aporia" I am intending to invoke its literal sense of "without passage" or "without *complete* passage" – what the Hamlet-scholar Stephen Dedalus might resignedly acknowledge (in Ulysses) as a "disappointed bridge."

across? What in science is without passage? Too often the crisis in the humanities is understood to pertain to the abyss between them and the sciences; but the abyss, like the crisis, is elsewhere – within, not between. Bridgework is accordingly no less necessary in the sciences than in the humanities. But what the former understands as a simple bridging problem, solved routinely every day, the other recognizes as the difficulty of getting from where we are, which is, as yet, nowhere, to the far bank.[5] In other words, if ours is a bastard age, it is so because it builds bridges without noticing that they fail – without noticing their gaps and the ways in which what they gather together they also disperse and keep apart.[6] We live in a metaphoric or almost metaphoric age having lost our awe of metaphor[7] – and this is nowhere more evident than in the overdetermined neglect by the sciences of their own rhetoric and language. That neglect *is* the crisis of modern science – except, paradoxically, it isn't. The crisis doesn't appear. Much of this book is an attempt to throw a sheet over it, to force it into appearance even if in the mode of hiding, like a cat under the bed covers.

But it's always a particular cat who hides under the comforter, not catness as such. So let me introduce a clowder

---

[5] "There is first of all the problem of the opening, namely, how to get from where we are, which is, as yet, nowhere, to the far bank. It is a simple bridging problem... People solve such problems every day..." J. M. Coetzee, Elizabeth Costello (2003), page 1.

[6] I am playing with Heidegger's discussion of bridges in "Building Dwelling Thinking," in particular his claim that "the bridge *gathers* the earth as landscape around the stream." In Heidegger's Basic Writings.

[7] < μετα + φερειν, a translation (a having carried across) of which is *to carry across*. "Almost metaphoric" since passage is incomplete.

of cats. In September of 2015, the internationally prestigious scientific journal *Nature* dedicated a special issue to interdisciplinarity,[8] in part to emphasize the enormously complex problems which face humanity and which no discipline alone can address. Yet *Nature*'s conception of interdisciplinarity was shockingly narrow and barely including the humanities.[9] Moreover, when it did, it was in a telling, instrumentalizing form.[10] Planetary salvation – as we shall see,

---

[8] *Nature* 525, September 17, 2015.

[9] The humanities were occasionally mentioned in the issue, including in its introduction (Anonymous 2015a), which described collaborations among "scientists, curators and humanities scholars" and cited in support a later article whose 'rich' defense of the non-mathematical liberal arts began with Steve Jobs and a calligraphy class (see below). A second article, albeit one dedicated to discussing global funding priorities, even managed to give an example of a science-humanities collaboration – a neuroimaging study of brain function during the reading of a poem – but failed to discuss the contribution of the humanities to that project beyond the poem's procurement (Rylance 2015). These rare mentions aside, the issue almost immediately forgot the humanities. Thus, the same opening introduction that mentioned "humanities scholars" was subtitled "Scientists must work together to save the world." A later essay was simply called "Team Science" (Ledford 2015). Likewise, the editors, in their announcement of the special issue, dedicated it simply to "interdisciplinary science" (Editors 2015). An article specifically focused on catalyzing collaboration discussed it in terms of cooperation only between biophysical and social sciences (Brown, Deletic and Wong 2015). On this note, let it be observed that the latter are often treated only marginally better than the humanities and frequently are reduced to an ancillary status charged merely with, e.g., finding "minimal, non-disruptive ways of dealing with social and ethical issues" (Viseu 2015).

[10] As an example of the merit of the humanities and the (non-mathematical) liberal arts, it called attention to a calligraphy class that apparently inspired Steve Jobs "to promote proportional fonts and establish Apple as the gold standard in desktop publishing" (Pormann, 2015). Thus the humanities *could*

the curiously apocalyptic backdrop of progress to which the special issue repeatedly referred – depended on the sciences working together, yes, but on the sciences alone. Meanwhile, in our pounce – if it isn't a destruction – of cats are two ostensibly scientific projects wielded in the service of an unexamined humanism: the exultant attempts to "de-extinct" lost animals,[11] on the one hand, and on the other, the triumphant synthesis of a so-called "freely living" cell with a supposedly minimal genome[12] – each of them the wondrous accomplishments of a science whose power is taken as an index of its progress. Yet, as I shall explain (especially in Chapters 1 and 3), both of these projects are built upon and work only within, not only a remarkable philosophical blindness, but a related *empirical* one as well. Notably each also explicitly operates in the shadow of crisis which, as before in the *Nature* issue, each would simply and heroically resolve.[13] If crisis is, as

---

profitably be put to use, but their value certainly did not entail questioning prevailing notions of value. Rather, it was only within those prevailing notions that they could, in an extremely limited way, be appreciated. The role of the humanities in critiquing foundations, means and ends was correspondingly as conspicuous for its absence, on *Nature*'s account, as was any such critique itself.

[11] For examples, consider Sherkow and Greely (2013), Zimmer (2013) and http://fivethirtyeight.com/features/the-scientist-trying-to-reverse-extinctions-think-jurassic-park/.

[12] E.g., Hutchison et al. (2016a,b), Service (2016) and Twilley (2016).

[13] De-extinction would supposedly mitigate the ongoing mass extinction attributable at least in part to human activities, while the creators of synthetic cells have often discussed them as possible solutions to health and energy crises. For example, see Thomas Goetz's interview with Craig Venter, a leader in synthetic cell science, published in *Wired – New York* (2012, p. 108) and titled "Life Hacker – He sequenced the human genomes and built a

I shall argue, the horizon surrounding all human activity, a projection of our own absolute and inescapable finitude or, in the language of Heidegger, of our own most potentiality for being, these projects stand out for their naïve if endearing faith first that the crisis is outside and not within, and second, that that horizon can be reached and then surpassed – that we can, as it were, put it behind us.

Knowledge would be the tool of that overtaking, that catching up with and getting in front of a receding, external horizon. But this knowledge is of an instructively peculiar nature. It is a knowledge understood as reducible to information, as bite-sized and modular and divorceable from any specific context. Such a view of knowledge is prevalent in the culture at large, including in the modern conception of the university as an information dispensary and of the teacher as information's apothecary, as well as in a scientific and scholarly literature which dribbles knowledge in ever-accumulating increments neatly fitted to the required brevity of an abstract. Knowledge in our era would be individuated and autonomous, yet paradoxically also disembodied. It is acquired by smartphone from a website with as much legitimacy as by painstaking inference, since the bottom-line, the information that knowledge is, will be true whether or not the accounting is understood. This individuation and autonomy of knowledge nonetheless peculiarly disembodied from context conforms to its project: it would be the vehicle of a parallel, human disembodiment, by which our horizons, our limitations and our finitude – in short our contexts – would be

new organism from scratch. Now Craig Venter wants to solve the world's energy crisis."

surpassed, even as we retained our autonomous and individual identities. Knowledge so understood is thus the mirror of the desire of which it would also be the medium. The corresponding relief of the ostensible knower from the responsibility of judgment – of criticism in the etymologic sense which links it as well to its cousin "crisis"[14] – follows as a corollary: for that responsibility is incurred by the very limitation from which knowledge, conceived paradoxically as both completed and limitless, delivers us. Knowledge would thus liberate man from the finitude which has ever defined the arena of his freedom; knowledge, become an absolute determinism, happily frees man from the agony of choice.

The impoverishment of knowledge and forgetting of the humanities together correspond to a similar impoverishment and forgetting of the human.[15] The divide, taken as self-evident, between humanities and sciences finds its reflection in an equally self-evident divide between humans and nature. The first term of this latter pair does, it is conceded, merit scrutiny, but in a form which, as we shall see, is simultaneously aggrandizing and belittling. Yet the resultant caricature is in fact not a modern innovation: it intimately evokes, or merely repeats, or even is the culmination of, a Cartesian gesture bifurcating the world into thinking and material things. The first of these (with which Descartes identified the human) being immediately apprehended, only the second required methodical investigation of the kind that became science. Thus man, as thinking substance, was expressly akin to the divine while being, as material thing, simultaneously overlookable;

---

[14] Both words are related to the Greek κρίνειν, to judge or decide.

[15] A forgetting of what we perhaps have yet to know in the first place.

though god-like, we might still confuse him with an automaton in a raincoat. Human forgettability thus became the ironic correlate of human exceptionality.

The *Nature* special issue depicted this irony in an especially interpretable, if not particularly sophisticated, way. In addition to the glorifying terms in which scientists were described – "Scientists must work together to save the world" read the subtitle of one contribution (Anonymous 2015a) – the issue's articles were complemented by cartoons portraying human beings – and specifically scientists[16] – as either superheroes or gods, dressed in skin-tight costumes, wearing masks and often flying. One of the cartoons was in fact used as a tag to identify all of the articles in the issue pertinent to the subject of interdisciplinarity.

On the one hand, then, in the mode of aggrandizement, scientists are like Thor, Wonder Woman, Superman, Captain America or Iron Man: divine, semi-divine or quasi-divine beings who have overcome their all-too-human limitations of finitude, mediation and materiality and from the vantage of their accomplished plenitude are thereby entrusted with planetary salvation. Knowledge would here be the means to superhuman extension and exaltation, and indeed, far from coincidentally, it is, per the comic book storylines, technological science which augments men (Captain America, Iron Man) into the ranks of gods (Thor, Wonder Woman) or god-like aliens (Superman). On the other hand, however, in the mode of belittlement, scientists are also like Thor, Wonder Woman, Superman, Captain America or Iron Man: comic-

[16] That images were specifically depicting scientists is suggested by the fact that one of them illustrates the title page of the article "Team Science."

book characters and icons of mediation, culled from a world of juvenile fantasy and computer-generated cinematic special effects and instrumentally mobilized as part of global marketing campaigns. Their great feats of superhuman strength are illusions, in two senses: the illusions are themselves remarkable feats; and the remarkable feats they simulate are illusory. Thus are god-like aspirations attained – in caricature – and the human being aggrandized as divine is here forgotten as fantasy.

Criticizing such a project, when its own autocriticism tags along like a clumsy alter-ego, might accordingly seem unnecessary. Nothing could be more false. Reading these cartoons is not a task that the sciences undertake for themselves. Instead, they typically understand these and similar features as little more than eye candy, merely accidental attributes of the coordinate articles and hence neither to be taken seriously nor meriting scrutiny. Likewise, the grand and aggrandizing claim that scientists are being called upon "to save the world" passes unquestioned – as also does any deep questioning of what such salvation would look like. Instead, the correspondence that *Nature* published in reaction to this special issue typically focused on discussing further ways for interdisciplinary research to proceed "successfully" – never once making explicit what "success" would really mean.[17] Yet the very examples offered of such successes, as well as of the problems modern science must solve, point to a much more complicated and ambiguous scientific role. Thus, the special issue editorial listed, of all things, the Manhattan project as an

[17] For examples, see Bammer (2015), Bursztyn and Purushothaman (2015), Szostak (2015) and Wassen and Heckert (2015).

example of "the best interdisciplinary science," as if the development of an atomic bomb were clearly a good in its own right and the participating scientists insulated from both the ostensible purposes and the still-unfurling consequences of the bomb's invention. Likewise, the "grand challenges facing society" to which the issue called particular attention were "energy, water, climate, food, [and] health."[18] Each of these is a problem which science must not only endeavor to solve or attenuate, but to the causes of which science historically has contributed and continues to contribute, often in major ways. Consider only the first example, energy: the scientific discovery (e.g., through geological prospecting), retrieval (e.g., through fracking or atom-splitting) and utilization (e.g., through nuclear power plants or the burning of fossil fuels) of various forms of energy are of course inseparable from major problems, from global warming to Fukushima, affecting water, climate, food and health. In fact, all of the challenges listed by the *Nature* issue are inter-related, with the scientific role *always* being two-sided, as both intended solution and unintended cause, as both publicized remedy and concealed poison – much like the superheroes already mentioned, who in saving the world destroy it and who thus become the threat they must protect the world from.[19]

---

[18] Anonymous (2015a). Note that these challenges are not otherwise described than by these common nouns, these synecdoches. One of the cartoons mentioned above is slightly more explicit: it lists, from left to right, "water shortage," "energy crisis," "disease" and "climate change."

[19] Goethe already thematized this problem when he had Faust lament that he and his father were praised for curing an epidemic though "... with our infernal tonic / Upon these hills, these dales we visited / A plague far worse than the bubonic. / Why with this poison I myself defrauded / Men by the

Whereas the comic book and film industry have thematized the equivocal role of the superhero, science typically has lacked that degree of sophistication, instead being prone to dismiss the mistakes of the past as the consequences of inadequate knowledge while promising only a single-edged sword in the future. Its past pretensions to plenitude and immediacy, though failed, in no way jeopardize its renewed pretensions now.[20] Even when confronting the ambiguous legacies of those it claims as its own – such as the multidisciplinarian Richard Francis Burton, advocate of a racist form of polygenism and subject of one of the special issue's articles (Pettitt 2015) – it is usually as a cautionary exception rather than a paradigmatic example. Moreover a cautionary exception from which no meaningful conclusion is drawn: thus the special issue's throw-away conclusion, immediately after citing the article on Burton, that "Today's would-be interdisciplinary scientists can draw many lessons from those of the past and can take our polymathy quiz online" (Anonymous 2015a). What an interesting pairing: a quiz which half-seriously pretends that polymathy is really

---

thousands, leaving them for dead; / Now I must hear the brazen killers lauded." (*Faust*, part 1, lines 1050 – 5, translated by Arndt).

[20] So does Nathaniel Hawthorne paint his "man of science," Aylmer, in "The Birthmark," a short story which richly anticipates many of the arguments and problems of my book. In this case, even though Aylmer's meticulous laboratory journal revealed "that his most splendid successes were almost invariably failures, if compared with the idea at which he aimed," he nonetheless did not hesitate to sacrifice his wife at the aesthetic altar of his science: "'The concoction of the draught has been perfect,' said he, in answer to [his wife] Georgiana's look. 'Unless all my science have deceived me, it cannot fail.'"

reducible to, and hence revealable through, multiple choice questions, and the idle invocation of unstated lessons from a little-known and less-examined past. Meanwhile, besides the more subtle ontological and ethical questions at stake in interdisciplinarity, the epistemological questions – e.g., on what basis do we accept as knowledge the purported syntheses born of massive interdisciplinary collaborations? – passed unremarked while instead the special issue concerned itself primarily with such important problems as credit dispensation, funding prospects and career advancement.

In later chapters (especially the third and the fourth), I try to discuss this not as some modern anomaly representing the perversion of proper science, but as traceable ultimately to science's deep roots in the endeavors of Bacon and Descartes – and, even further back, perhaps surprisingly, of Aristotle. In Bacon we see the project that became science cast in explicitly restorative and Christian terms, as man's venture to reclaim both a dominion and a divine likeness lost with the Biblical fall. That its modern realization should literally be cartoonish speaks as much to the impoverishment of prevailing views of the divine as of the human. In a different (but related) vein than Bacon's, Descartes also cast his project in explicitly theological terms. As already noted, he separated man's pretensions to the divine from his material limitations through the stark dualism he inaugurated between the incorporeal thinking subject and the extended physical world. Modern science may think it has cast off the theological and philosophical prejudices of its progenitors, but that casting off, I argue, is little more than a cultivated obliviousness. Moreover, as radical as the Baconian and Cartesian developments might seem in the context of the tradition from which they appear to break, I try to show instead their deeper continuity with that tradition, particularly in my analysis (in Chapter 4) of Aristotle's conception of the human desire to know. What links all of these cases is man's present – but, they think, wholly accidental and hence unnecessary – inadequacy. I attempt instead to read the necessity of that accident. I also attempt to show how such readings can be pursued with technical scientific texts themselves – that is, the ways in which already legible in these texts is their own critique, much like those accompanying cartoons of the *Nature* special issue. I do this especially with a detailed reading of Darwin's On the Origin of Species (in Chapter 2, which focuses on the disjunct between the ontology he assumes and the one he describes) and of several papers focused on molecular biology (Chapter 1, examining the deconstruction of "self" and "essence" in these texts) and conservation biology (Chapter 3, about the attempts to distinguish ecological corruption and restoration).[21]

I want especially to emphasize the importance of bringing this style of criticism to scientific texts, not because others have not attempted it but because the field of science remains stubbornly insulated from any but the most naïve theories of reading – and it is according to those naïve theories that scientists typically read each other's work (to the extent that they read them at all; see Chapter 4). Someone must read scientific works better. In fact, a science to come – a post-

[21] Thus, in terms of the disciplinary frame of the modern academy, Chapter 1 deals with molecular biology, Chapter 2 with evolutionary biology, and Chapter 3 with conservation biology. As will become clear over the course of this book, however, such neat divisions are illusory.

humanistic science – must learn to read altogether differently; must learn, if nothing else, that reading has never been self-evident; must learn in turn that, if reading is not self-evident, even less so is writing. A science to come must confront how, far from its wielding writing, it is writing which wields it. But this means recognizing that writing is absolutely not a transparent medium by which information is simply transported from here to there.[22] It means recognizing an aspiring bridge has managed only to be at best a rather short pier. Scientific writing, with its frequent passive voice and storied distaste for rhetoric,[23] may seem the exception, due to its rigorous regulations and policing. Those are characteristics, however, not of scientific writing, but of scientific reading – in fact, of poor reading. I try to show instead that careful attention to scientific texts themselves reveals something richer, if largely unread. Often at stake in scientific texts, in their form as well as in their content, from Darwin to the recent technical literature, is a desire for autonomy, self-sufficiency, wholeness – for simple, unmediated presence – frequently projected onto the scientific object (for example, the living organism), but, even more so, onto the assumed character of the scientific knower. Ontology and epistemology thereby become confused and, if in a distorting way, mirror one another – as do the initial, more ontologically-focused

[22] As implied earlier, such an image ironically underlies the etymologic roots of both "metaphor" and "translation" – the first being a carrying across and the second its Latinate and now accomplished (perfect tense) passage. I discuss this further in Chapter 2.

[23] Ironically, its deployment of the passive voice is itself a rhetorical gesture signaling a promised neutral objectivity.

chapters (1 and 2) of this book, and the later more epistemologically-focused ones (3 and 4). What is known and how it is known, it seems trivial to remark, are not unrelated; but their relationship, it turns out, is far from trivial. Throughout this book, I read both as expressive of that metaphysical desire for unmediated presence mentioned above, a desire which excuses if it does not justify the philosophical priority bestowed upon essences, substances, subjects and individuals over relations, contexts and, for want of a better word, ecologies.[24]

To give but one example of such metaphysical commitments, consider the extraordinary role accorded to DNA, to which we turn to explain everything from criminality to sexuality, as if this seemingly endogenous chemical, not to mention the traits thought reducible to it, were in nowise influenced by the myriad contexts, whether social, ecological or epigenetic, in which it is expressed. This unrealistic, hyper-explanatory role comes to the fore in the projects I've already mentioned, de-extinction and synthetic biology. Thus, according to paleontologist and self-professed species revivalist Michael Archer, "We've come a long way in our understanding about what makes animals and what makes plants. We know it's essentially their DNA. We now understand how DNA

[24] And yet, if this looks like a mere opposition, it is not: the terms on each side of the divide mutually determine one another, in a sort of divergent dialectic one expression of which (discussed already and returned to especially in Chapter 4) is the simultaneous aggrandizement (as individuated, god-like knower) and belittlement (as mere cog in an enormous relational enterprise of knowing) of human beings. Rather than choosing sides, this book is endeavoring to understand whence the polarity emerges.

works. We now understand how it expresses and develops creatures." Later, he continued: "If we can get the thylacine's ancient DNA back into an embryo, it will produce a thylacine."[25] Such an assertion takes for granted that we know what a thylacine is and that it's basically nothing more than a form, a shape, for which its DNA is code.[26] But is that truly what a thylacine is? As I try to show in the second chapter, Darwin, despite the privilege he gave to inheritance (understood as success in leaving progeny), might nonetheless argue otherwise, calling attention instead to the infinitely complex mutual relations of organic beings one with another. He might ask, is something that looks like a thylacine and has the DNA we call a thylacine's still a thylacine outside of an ecological context which includes, for example, other thylacines, as well as prey, predators and competitors, with all of which thylacines once interacted? This is to say nothing of microorganisms and the corresponding thylacine microbiomes, whose important roles can range from disease resistance and digestion to development and behavior.[27]

---

[25] The quotation is from the 8-minute documentary "The case for de-extinction" produced by ESPN films as part of the Collector Series. The first quotation begins around the fifty-third second, while the second starts around five minutes thirty seconds. Thylacines are extinct marsupials. See http://fivethirtyeight.com/features/the-scientist-trying-to-reverse-extinctions-think-jurassic-park/.

[26] Needless to say, it also takes for granted our own self-evidence, the transparency of the "we" wielding its unquestionable knowledge.

[27] Such critiques have in one form or another been raised. John Rennie, the former editor-in-chief of *Scientific American*, observed in the afore-cited video that "the real ethical irony of de-extinction is the idea that we would be able to try to help make nature more natural than it was, we'd be able to return it to a more pristine state. The problem is that in practice a lot of the time we

According to what mad desire, then, would we reduce an organism "essentially" to its DNA, while in parallel dismissing these other "infinitely complex" relations as mere accidents irrelevant to an understanding of what thylacines were or could be? With what mad desire do we understand living beings, including ourselves, as the elaborations only of an internal, self-constituting directive and not equally the imposition or impressment of a network of relations, both internal and external, with others?

In calling this a "mad desire," I am also suggesting that, despite universal pretensions otherwise, and despite the corresponding exaltation of an unmediated and hence fully present subject, projects such as de-extinction are not in fact

---

would end up creating a kind of simulation of an extinct creature and we would be putting it back into a world where it wouldn't naturally have a place." This is good as far as it goes; but the real work begins when one interrogates what could possibly be at stake, first in making nature even more natural, and second in the notion of a creature – or rather a simulation – "naturally hav[ing] a place." On the first, see Chapter 3, as well as Nathaniel Hawthorne's "The Birthmark," whose "man of science" enjoins his wife to "doubt not my power," as he anticipates his "triumph when I shall have corrected what Nature left imperfect in her fairest work!" Never does he question whence derive his own aesthetic commitments. The second, meanwhile, will be a principal topic especially of Chapter 2.

Likewise Sherkow and Greely (2013) stopped where they should have started when they asked, "Does one individual (or a set of clones) make a "species"? Even if genomic identity is necessary, is it sufficient? The revived individuals would not have the same epigenetic makeup, microbiome, environment, or even 'culture' as their extinct predecessors." Good questions – but they never endeavor to answer them, much less to consider their implications for normative understandings of organisms, individuals, species or even the entire projects of biology and ecology – the subjects of Chapters 1 and 2, especially.

founded on knowledge, as I explain in Chapters 3 and 4. The claim that we know how DNA "expresses and develops creatures" is recklessly ignorant not only of the pertinent ecology (as already noted) but of the pertinent molecular biology. Nor are those two clearly separated, since, for example, studies of microbiomes suggest that biological development may reflect ecological interactions. Further, were it truly the case that we had mastered such knowledge, the effort to clone the extinct bucardo (also known as the Pyrenean ibex) would not, out of hundreds of attempts, have resulted only in a single fetus coming to term – and dying within minutes of birth, apparently from respiratory failure associated with major lung abnormalities.[28] This is to say nothing of a project of knowledge which remains oblivious to its own metaphysical dimensions, dimensions that not only privilege essence over relations but do so within an explicitly Christian rhetoric. For it is not nothing that the name of one of the endeavors Michael Archer heads is the "Lazarus Project," as it is not nothing that frequently de-extinction is described in terms of "resurrection."[29] Archer's own defense of species

[28] Folch et al. (2009). To be clear: had the cloning succeeded, that would not necessarily imply our mastery of the relevant knowledge. There is much that one can do without knowing what one is doing; procreation is and has always been high on that list. Of course we would also have to ask what mastery of knowledge could possibly mean, and where the limits of relevance are.

[29] Thus, for example, Michael Archer's March 2013 TED talk was titled "How we'll resurrect the gastric brooding frog, the Tasmanian tiger." See also Saul (2013) and Australian Associated Press (2013). Similar rhetoric is also used to describe related conservation projects like rewilding, as pointed out in Chapter 3.

revivalism is testimony enough: "If we destroyed part of Eden," he argued, "we are responsible for fixing up that garden. This idea of restoring things that have become extinct has actually a Biblical sanction. 1 Corinthians 15 verse 26 says 'the last enemy that shall be destroyed is death.'"[30] Far from seeing such an explicitly Christian justification as anomalous, it behooves us to recognize its extreme continuity with the scientific project from its early articulations by both Bacon and Descartes. As already noted, Bacon in particular understood his proposal as a restoration consequent upon man's fall. As I argue in particular in the third chapter, this provides not only a theological frame but elucidates how the acritical privileging of essences is founded on a transcendental monotheism whose achievement would in turn be a transcendental signified.

Many of the same problems come to light in a close examination of synthetic biology's claimed synthesis of a "freely living cell" with the smallest known genome. As with de-extinction, this project begins from an exalted and essentializing conception of DNA, moreover one elaborated upon specifically by analogy to computers: "The genome sequence of a cell may be thought of as its operating system," Hutchison et al. (2016a) write. A few sentences later they continue: "The genome can be viewed as a piece of software;

---

[30] The quotation is again from the 8-minute documentary already cited. It begins at approximately 6 minutes 20 seconds. Note the immediate confusion of signifier for signified: as if bringing back an image, ideal or simulacrum of the dead is equivalent to bringing back the dead and hence to conquering death. This imposition of an image to be extolled as what it represents is a common feature of conservation projects, as Chapter 3 endeavors to demonstrate – a gesture which, instead of vanquishing death, is conceivable only on the condition of death.

DNA sequencing allows the software to be read."[31] Then, after observing that "deciphering the operating system of the cell was a daunting task," they announce their goal: "We have long been interested in simplifying the genomic software of a bacterial cell by eliminating genes that are nonessential for cell growth under ideal conditions in the laboratory" – a goal they believe themselves to have achieved. No doubt their recourse to computer imagery is as natural to them as Archer's was to resurrection, all the more because similar imagery has pervaded molecular biology more or less since the discovery of the double helical structure of DNA and the framing of its interpretation as a problem of coding, whether a computer's or Morse's.[32] But we could trace the roots of such imagery much further, at least to Descartes and his reduction of the physical body to automata or machines. That reduction, for Descartes *and for his heirs*, had as its corollary the liberation of the cognitive, spiritual being from its material husk. The disembodied knower, whether in the 17th or the 21st century, confronts the biological machine on the postulate of which he piggy-backed his own disembodiment.

In its modern articulation, the problem folds on itself: a human perspective that is implicitly disembodied and acontextual carries as its correlate an organism – or, for that matter, a gene – which, though understood as a machine, is

---

[31] Undoubtedly they do not realize that they have already postulated DNA as a sort of palimpsest, readable at one level by human beings using sequencing technology, and readable at a different level, in a different way, and with a different message, by, e.g., cellular enzymes.

[32] Crick (1954, p. 60) organizes the problem of DNA interpretation by analogy to Morse code.

simultaneously essentialized outside of all context. Hence is born, in parallel to or parody of the envisioned freedom of the human investigator, the notion of a "freely living" organism (Service 2016) or of "an autonomously replicating cell" (Hutchison et al. 2016b) – even if that organism is a derived parasite growable only in rich, highly defined culture media (as in fact is the case for the synthetic cell achieved by Hutchison et al.). The provision of that environment by other living beings – in this case scientists – in no way jeopardizes the organism's independence or autonomy, though the same would not be said were it found in a more classically parasitical context. Indeed, the conclusion nearly diametrically opposite to the one I am suggesting is reached: precisely *because* the medium "supplies virtually all the small molecules required for life," the genome enabled therein is understood to "reveal a core set of environment-independent functions that are necessary and sufficient for life" (Hutchison et al. 2016a). The "ideal conditions in the laboratory" have been mistaken as literally ideal, as an environment that isn't really there. Where this logic leads is monstrous; I draw the inevitable conclusions in Chapter 1. Among other consequences, despite its reliance on a hidden essentialism – here invoked twice, first by the designation of some genes as essential or non-essential, and second by the absurd assumption that indispensability (taken as a proxy for essentiality) is a characteristic solely of the gene and not of the context – this essence-privileging logic nonetheless cannot help but critique essences. This is already legible in Hutchison et al.'s invention of a third category, the quasi-essential,[33] as well as in the tortured logic by which they arrive at nonessential genes with essential biochemical functions.[34] When essences become quasi or conditional, or when they come in degrees, what truly remains of them? Under these circumstances, so-called "freely-living" organisms are revealed as no less reliant on their biological neighbors as the obligate symbiont or parasite, though in a more subtle way. Put more strongly, we are forced to admit that there is not, has never been, nor ever will be such a living being, whether thylacine, microbe or man, as "freely lives" outside of

---

33 Genes which caused "growth impairments" or "slow growth" were classified as quasi-essential, without pausing over two major difficulties: first, how do you distinguish no growth from extremely slow growth? That is, how do you distinguish an "essential" gene whose removal prohibits growth absolutely from a "quasi-essential" gene whose removal permits growth only over some, to our perspective, unduly extended time period? (It is noteworthy how temporality recurs over and over as a problem, whether in the pursuit of essential genes and biological origins [Chapter 1], Darwin's articulation of his evolutionary theory [Chapter 2], conservation biology [Chapter 3], or the pursuit of knowledge and concomitant understanding of authorship [Chapter 4]. Its recurrence must not be brushed away through a naïve invocation of historicism. It instead must be examined, not in terms of the rhetoric of temporality but in terms of temporality as rhetoric. Chapter 1 discusses why.) And second, is this new category of the quasi-essential not a covert acknowledgement that, though sluggish growth is indeed tolerated in the specialized environment of the laboratory, it would not endure in many other environments to which the experimenters, despite pretensions to the contrary, nonetheless desire to make claim? In other words, is it not covert acknowledgement that these "functions" are anything but "environment-independent"?

34 "In a few cases, n-genes [= genes which they have designated nonessential] were retained – specifically if their biochemical function appeared essential" (Hutchison et al. 2016a).

any dependency on context, which latter is always inseparably both biotic and physical. Moreover, as I argue especially in the first chapter, this inescapable dependence on context undoes, both in theory and in practice, any essential distinction, however desired, just as it undoes the organism's desired autonomy, which would be but the projection of a disembodied cogito onto its machine plaything and secret likeness.

Yet, far from recognizing the dissolution of its desired categories, the project of synthesizing life, much like that of de-extincting organisms, proceeds apace. The disembodied and unmediated cogito that would master the material automata which, in its eyes, constitute the world can do so on the sole condition that it be by its own design. There is no mastery without design or purpose.[35] Thus, the synthesis of life counts as man's achievement only if it is done at man's instruction; and this is possible only if the living entity, alongside its human author, can be rid of all of its other dependencies. If instead that being lacks an essence, if no "essential" genes can be found for it, if no characteristic can be linked essentially to it, then the human attainment of god-like powers to create or

[35] This is a further, covert accomplishment of Descartes: by reducing the human being to a material thing *of a particular kind* – an automaton – Descartes preserved, through that postulated instrumentality, a higher end, an author or knower or thinker who, if nothing else, operates the machine. This *res extensa* requires a *res cogitans*, even if their interaction, or even the possibility of their interaction, remains mysterious (as Elizabeth of Bohemia obstinately and rightly pointed out in her letters to Descartes). (That animals are also automata is no problem for Descartes, since humans instrumentalize animals.)

resurrect life[36] will itself be determined by an unmastered and unmasterable context and will thereby be revealed as contingent. In turn that contingency can only infect our purposes, both because we cannot impose a steadfast purpose on an always contextual being and because that purpose itself is contextual and hence not steadfast. Absent enduring essences, enduring purposes are also absent. Thus, despite itself, the disembodied cogito finds itself clothed in bodily raiment, its pretension to universality no less contextual and contingent than the essences and purposes upon which it relied. The cogito's deliverance *from* limitation is exposed as a defiant but ineffectual protest *to* limitation, and the illusion of that deliverance, like the superhero illusions of contemporary comic books and films, can only testify to the overwhelming strength of a desire.

Such a desire is expressed in all of these projects, from the embrace of interdisciplinarity that accepts without question the human/nature divide and in parallel excludes the humanities from its synthesis, to the nature-mastering and death-conquering projects of resurrecting life or creating it by design. All are united by an unattainable desire, the backside of a metaphysical foundation which gives these projects their seeming coherence. I have discussed this so far (in sketch only), as well as (in detail) in the chapters to come, in terms of Bacon's Biblically inspired project of restoring human mastery ultimately over the universe; and in terms of the radical dualism by which Descartes constituted a natural world as the province of study for a disembodied cogito. But here Bacon

[36] How can the Judeo-Christian dimensions of these projects, creating life and resurrecting it, be so regularly overlooked?

and Descartes are best understood as figures, similar to those variables which Descartes deployed in analytical geometry in order to name the unknown and thereby to begin to know it. We might ask what these men are figures of – but in so asking we risk repeating their error. (It may be that their error can only be repeated.) The challenge before us is to endeavor to think – or rather, to read – figures whose significations cannot come to rest in a final signified. The assumption, ultimately perhaps unavoidable, of that non-signifying signified, fully present and removed from all mediation and context, is what I refer to as the metaphysics of presence – and it is this assumption which I endeavor in each of my chapters to expose and challenge tenaciously.

My use of this term is guided by Derrida's critique in Of Grammatology of the disparagement of writing found throughout the western philosophical tradition. What is threatening about writing? In contrast to the supposed immediacy and presence of the 'living voice,' the voice which can guard its verbal offspring and clarify misunderstandings and which is simultaneously the breath of life, writing is detachable from the, again supposed, guarantor and back-stop of meaning which is its author. That detachability is not an accidental feature of writing but the condition of its possibility, mirroring as it does the detachability of writing from what would be its signified: for that signified, it turns out, can always be understood instead as a signifier. Writing, for example, is often taken to refer to (spoken) words, not things. Its signified would hence be a signifier. But is the spoken word's reference to things any more stable than the written words reference to speech? In ancient Egypt, monuments and statues were readable as hieroglyphic signs, even if they also were inscribed with such signs and referred to by such signs[37] – but this possibility, far from being an innovation of the Egyptians reflecting the peculiarity of their script, is in fact just another version of the same condition of possibility of writing. To declare as much is to admit that the confusion of signified and signifier, the two-way "perversion" of one into the other, associated with writing because writing is inconceivable without it, can nonetheless not be restricted to writing; or rather, it is to admit that what falls within the scope of writing must be radically expanded and rethought. The attempted ostracism of writing, so characteristic and simultaneously ambivalent a gesture in philosophy since Socrates cast his sherd in Plato's *Phaedrus*, would banish the condition of its own possibility. It is also, naïvely, to assume that writing is everywhere recognized as writing: as if the very circumstance which writing is always revealing, the confusion of signifieds and signifiers, didn't preclude absolutely the possibility that writing will simply, modestly and self-evidently declare itself.

---

[37] As recognized already by Hegel in The Philosophy of History pp. 214-5: "The force which urges [the Egyptian spirit] is the impulse of self-comprehension; and it has no other material or ground to work on, in order to teach itself what it is – to realize itself for itself – than its working out its thoughts in stone; and what it engraves on the stone are its enigmas – these hieroglyphs. They are of two kinds – hieroglyphs *proper*, designed rather to express language, and having reference to subjective conception; and a class of hieroglyphs of a different kind, viz. those enormous masses of architecture and sculpture with which Egypt is covered." Of course this leads one to wonder how a *proper* hieroglyph ultimately is to be distinguished.

These claims amount to the admission that the threat to the western philosophical tradition posed by writing cannot be contained or quarantined; speech, for example, is on this view writing. The security, the pledge, represented by the speaker is illusory, and she can no more command an end to the play of signification based on her original intent than she can comprehensively know that original intent in the first place. The so-called living voice is not immediate and therefore not immediately present, even if its mediating offset is nothing more than an instant of life-betraying breath. This in some ways was the insight of Freud: he read the (often spoken) words of others and found thereby the coherent expression of what he interpreted as intentions, though they were intentions unknown to the speaker. And it was no less his mistake, both in recuperating meaning to an intention that, however hidden, remained in principle knowable and, relatedly, in endeavoring to define an interpretative lexicon that determined absolutely the parameters of signification's play. Thus the very Freud who recognized the possible estrangement between speech and speaker recuperated that estrangement via a retrievable and final meaning. His critique of what I am calling the metaphysics of presence, his willingness to demonstrate repeatedly that such-and-such a signified was itself signifying, ultimately re-arrived nonetheless at that same metaphysics, at a final signified at last removed from the perpetual play of signification.

Avoiding that return is not a mere matter of vigilance or rigor. If writing is understood in the expansive sense being suggested, then it isn't just the distinction between signified and signifier that is destabilized. No category of thought, including "category" and "thought," can be insulated from the implications. Not truth, not knowledge, not the good, not being, not becoming, not subject, not object and not god. Yet, to pretend, Descartes-like, that we can somehow rid ourselves of these orienting concepts and start from *tabula rasa*, having doubted all that is doubtable – is precisely again to fall into the metaphysics of presence by implicitly assuming a position outside of context, hence unmediated, hence unsignifying. That Descartes failed to find such a position is already implied by the first person singular active voice that he somehow smuggled into his undoubtable proposition *cogito ergo sum*. If nothing else, Descartes could not rid himself of grammar – one of the many contexts into which we are thrown inextricably. Our project, then, cannot be to identify a secure spot from which to begin anew, an origin, intention or meaning that we would by agreement, pretense or sheer desire define as impregnable. Rather it can only be to read, necessarily from within the metaphysics in which we are situated and which situates us, against the grain and thereby to attend the irreducible differences between what would be *said* from within the metaphysics we inhabit and what is *written*.

That in summary is the work I have endeavored to undertake, no doubt without final success, in this book. Whether reading what are more easily identifiable as texts, the articles and books associated with specific authors, or the no less textual phenomena of the world, and in particular of the living world, I have attempted to remain attentive to their graphic character. This means both their *signifying* character and their *material*, hence *limited* and *finite* character. Were it otherwise, were they immaterial, unlimited and infinite, context could, at least in principle, be dispensed with, and the flight of meaning which is always both its excess and its deficiency would find at last its precise commensurability with an origin and an end. Its full presence could finally be assured. This assurance is the postulate from which modern science proceeds, and it is its deepest continuity with a philosophical tradition of which it is largely and often willfully ignorant. But the assurance can only fail. Absent it, we arrive instead at the unintuitive conclusion that a world everywhere limited nonetheless has no determinate limits. Instead, context, which inescapably bounds, puts into play every provisional boundary, whether between self and other, subject and object, living and non-living, or signified and signifier.[38] Moreover, context, though inescapable,[39] is not inexchangeable, as the practice of excerpting or quoting trivially proves. Hence the determinate borders that are the characteristic of final answers have at best a performative nature.

Perhaps no scientific field reveals this as clearly as biology. The question "What is life?" posed now and again by scientific biology is not and has never been – despite the discipline's own self-understanding – answered by facts or by inferences from facts, but rather by fiat: and it behooves us to dwell on why only performative and legislative answers to such

---

[38] *Nature*'s multiple-choice polymathy quiz was in fact predicated on this inescapable, often invisible contextuality. Each question pertained to the understanding of a word whose meaning was equivocal and in particular depended on disciplinary context. In picking a specific answer – rather than all of them – you thus revealed your disciplinary parochialism. *Nature*'s own parochialism was in turn revealed by its incapacity to read the implications of its own semi-serious quiz.

[39] As I point out in Chapter 1, there is no outside to be outside in.

questions are possible.[40] It behooves us no less to bear in mind that such answers, like political regimes, change. The law can only ever announce its own possible transgression; the outlaw is born with it.[41] Thus, the implicit and oneiric answer to the question of the meaning of being – presence – can always be read as both the expression of a desire (want) and the confession of an absence (want). And so do I read it in this book, in its myriad appearances across the scientific literature from Darwin to the present day, including in the hollow pretensions of scientific authorship itself. We who, like Hamlet, have long been bound in a nutshell though counting ourselves kings of infinite space, must now attend to our bad dreams.

Hence the idiosyncratic title which I have given to this book. I do not offer a whole, but fragments; no paean of victory but songs of mourning. Both follow from my attempt

---

[40] I am hardly the first to make this point or comparison. N. W. Pirie implied something similar in his 1937 essay, "The meaninglessness of the terms life and living." There he noted that the absence of a "definable line between 'living' and 'non-living' matter" did not necessarily preclude the utility of these terms in certain contexts. To illustrate this claim, he drew an analogy to human perception of the electromagnetic spectrum: "If it should be decided that there is no definable line between 'living' and 'non-living' matter the terms will still be useful for most purposes, in the same way that the words green and yellow are useful. For, although it is obvious that so far as our unaided vision is concerned there is a smooth transition between these two colours, so little of our colour experience falls in the region 'greenish-yellow' that it is convenient to name regions of the spectrum as we do and to imagine that they touch along legally definable lines."

[41] That Francis Bacon was a jurist was not without consequences for the development of what has become modern science, including through his radical redefinition of form in terms of law.

to remain faithful to the very materiality that Descartes exempted man and god from; to that limitation and finitude which are everywhere boundless even as they are forgotten by the relentless march of progress they also enable. The thinking I am trying to invite is not a thinking from the other side, where crises have been met and challenges surmounted; not even a thinking which works its way to that other side when at last accidents have been banished and life's fragility impregnably buttressed. Such triumphalism already characterizes our thought, and in particular our scientific thought; it is a veiled form of essentialism and a discreet affirmation of presence; and its condition is blindness. What I offer instead would be a thinking from the end of the pier – from the limit of a failed bridge – from the far edge of the gangplank. It would be a thinking from within mourning, from within the certainty of absolute loss, from within fragments which can never unite into the purity of a whole.[42] In contrast, the thinking of our age appears daring, confident, unlimited and above all, possible. But this is only an appearance; and, as I endeavor to argue throughout this book, the thinking which we need must instead be everywhere limited and unconfident; daring in a new way which embraces responsibility rather than evades it; and as for its possibility – Is it possible? Is it impossible? Perhaps this would be a thinking which endures between the possible and the impossible, its endurance doing

[42] "Obviously," wrote Jorge Luis Borges in "John Wilkins' Analytical Language," "there is no classification of the universe that is not arbitrary and speculative. The reason is quite simple: we do not know what the universe is... We must go even further, and suspect that there is no universe in the organic, unifying sense of that ambitious word."

no more and no less than deferring the distinction between them. And so would it await the dawn.

First Episode (ἐπεισόδια): Knowledge of Life – Toward a Post-humanistic Science

Stasimon (στάσιμον): Proteus

His name is suggestive of prōtos (πρῶτος), "first," and if he is a god, he is also simply (and metonymically) "the old one of salt" (γέρων ἅλιος) or, less literally, "the old man of the sea."[1] But is he god or man or something else? What is he? Surely he must be a god, for he can change shape; and yet what does it mean to be an old shape-shifter? Is youth not also a shape? Anticipating our confusion about his own nature, he mocked us by naming his daughter Eidothea, "the image of the goddess." But we wish to get beyond images, whether his image or hers. We wish to get to what was first – setting aside the facts if we have to.

His kin was less than kind: his god-like daughter betrayed him to the Greek hero Menelaus. "He keeps, protects and loves a flock of seals," she told him. "If you wish for him to confide in you become a seal." But how does a man become a seal, Menelaus is said to have asked. "How would I know?" she answered. "I am but the image of the goddess."

"Ah!" thought Menelaus. "It isn't that I must become a seal; I must simply become the image of one." And so Menelaus killed a seal beloved by the old one, and he and his men ate the oily meat, and then he draped the hide over his

[1] It is also possible to translate the phrase as "the old, idle one" or "the old, fruitless one."

body, covering his nakedness, and lay on the rocks as he imagined a seal would. When the old one came ashore in order to rest – for, whether or not he was a god, he was old – he counted his flock, and among them he counted the image of a seal, which was also Menelaus and also a dead seal and perhaps was other things as well. If he was a god, Proteus was easily tricked; if merely old, perhaps his exhaustion contributed to his error. Or perhaps he made no error.

The old one then settled down among his beloved flock, one member of whom was dead, and drifted into sleep. "When my father naps," Eidothea had said, "seize him. He will change into innumerable different forms, of elements and of beasts, but do not let go of him; eventually he will tire; and when he returns to his resting form, if you have not let go of him, he will confide in you." But will he answer any question, Menelaus is said to have asked. "How would I know?" she answered. "I am but the image of the goddess."

So Menelaus waited until the old one – or if he was a god, the god – fell asleep; and then, still draped in the hide of a seal, he leapt forward and grasped at Proteus. But what did he grasp? Was it an old man? A lion? A snake? A leopard? A boar? A tree? Running water? A single licking flame? Air? A god? Something aged? An old man? And whatever it was that he grasped, what was it that he held onto?

Regardless of the answer, it tired before Menelaus tired; if it was a god, it tired or became bored before the man did; and if it was simply old, maybe oldness itself, it perhaps understandably tired of the game before this still-youthful ancient hero did. A voice called out from whatever it was that it was, and the voice said, "What would you know?"

Menelaus asked: "Is your resting form your original form?"

Proteus responded, "I cannot say."

Menelaus asked: "Is your original form your true form?"

Proteus responded, "I cannot say."

Menelaus asked: "Is your true form your final form?"

Proteus responded, "I cannot say."

Menelaus asked: "What is it that I have held onto?"

Proteus responded, "I cannot say."

Menelaus paused. "Can you at least say," he asked, "whether your daughter is a goddess or only the image of a goddess?"

"Even that," responded Proteus, "I cannot say."

"What then is it that you can say?" demanded the exasperated hero cloaked still in a seal's hide.

"If you wish, perhaps I can give an account (λόγος) of your life (βίος) away from your household (οἶκος)."

"And of my homecoming?"

The old one smiled. "My dear seal," it said, "are you not already home?"

Chapter One, on Molecular Biology
Grasping Proteus

"To become is not to attain a form (identification, imitation, Mimesis), but to find the zone of proximity, indiscernibility, or indifferentiation where one can no longer be distinguished from a *woman,* an *animal, or* a *molecule – neither imprecise nor general, but unforeseen and nonpreexistent, singularized out of a population rather than determined in a form. One can institute a zone of proximity with anything, on the condition that one creates the literary means for doing so."*

Deleuze (1998) "Literature and Life," p. 1.

I. Introduction

Since its birth, biology has been heir to one of philosophy's inaugurating questions: what is life? In biology, this question is, not unreasonably, heard only in some of its possible dimensions. For example, biology typically does not concern itself with an ontological examination of life's *being*; it does not ask whether life has a mode of being peculiar to it, different, say, from the mode of being of a rock or a word; nor, in any way does it ask whether its mode of being is being at all. For biology, the "is" in the question "what is life?" is not in question; life is. The biological approach to this question accordingly proceeds from the givenness of the "is." No less does it proceed from the givenness of "life." For the biologist, the question "What is life?" is to be answered by assembling a catalogue of living organisms and abstracting from it what characterizes them uniquely. In so doing, this Baconian procedure takes for granted one last aspect of the question. It accepts that life is a "what." Life, for biology, is a discrete, isolatable thing, something which can be catalogued in tables and atomized into traits – above all, something which can be defined, delimited, demarcated with a clear and distinct border which separates absolutely an inside from an outside.

In observing that biology hears the question "What is life?" only within a narrow range of possibilities, I am not endeavoring to scold biology, to mock it or to criticize it from the safety of the theoretical sidelines. I do not dispute that the way that biology has heard the question has been, for biology, very productive.[1] But I would like to propose that biology itself, for the most part unknowingly, has opened the door to other ways of hearing, or perhaps better, of reading, this question, ways which invite a different approach to the three words which constitute it. In the chapter on Darwin, I endeavor to work out, in a preliminary sense, some of the stakes implicit to the "is" of this question. In this chapter, I will instead focus on the "what" and the "life," leaving the "is," for now, bracketed, though it will be unavoidable that we nonetheless approach it.

As a *point de départ*, consider one of the more influential answers given to this question. Notably the answer is in the form of a definition, though qualified as only an operational one. That there may be other responses – perhaps not answers – to this question besides definition is not considered. Instead,

---

[1] One should note, however, that biology's understanding of its own productivity is not unrelated to how it has heard the question.

the influential biochemist and central figure in 20th and 21st century microbiology, Norm Pace, defined life as follows (2001): "Life is a self-replicating, evolving system expected to be based on organic chemistry." This definition, I should note, is the culmination of an explicit rumination on the question "What is life?" Such a solution is hardly unique to Pace,[2] and I will here treat it as illustrative of the prevailing approach to the question in modern biology.

Setting aside the expectation of organic chemistry – since the expectation is ours and not life's – I will focus on the other characteristics. First, life is self-replicating. By this one term, Pace meant two distinct things: first, that life makes copies of itself; second, that it itself makes the copies. Life powers its own replication; to be life, life must itself harness the energy and materials needed to make the needed copies of itself. In what follows, I will try to keep these different meanings both in view. I will ask, on the one hand, what counts as the "self" that life replicates? And, on the other, what counts as the "self" that does the replicating?

Second, life is, according to Pace, evolving. He imposed this criterion in part to exclude 'self-replicating' entities like crystals from being considered life. These systems, he observed,

---

[2] For example, NASA's Exobiology Discipline Working Group defined life as "a self-sustaining system capable of Darwinian evolution." While "self-sustaining" is not precisely equivalent to "self-replicating" – though the two are often taken to imply one another – the same arguments which I will make about self-replication apply to it, namely that a self sustained through and though evolving can only be a self not quite itself. Likewise, among historical criteria for life, adaptation, reproduction and growth, however murkily defined, are inevitably included, as Pirie (1937) observed – despite his refutation of each.

"do not evolve, do not change in genetic ways to meet new challenges." The implicit definition of evolution as "change in genetic ways to meet new challenges" is key; not because it emphasizes genetic change – clays, for example, have been interpreted to change in genetic ways (e.g., Cairns-Smith 1985)[3] – but because it emphasizes meeting new challenges. But what counts as a new challenge, one which clays could not meet? Here an interesting tension emerges, for those challenges are none other than "the necessarily competing drives for self-replication that are manifest in all organisms." Natural selection, which Pace identified as "the mechanism of evolution," is a consequence of self-replication. But it is a consequence that necessarily affects what self-replication means, in both the senses delineated above. For a self-replicator to evolve, it must replicate something different, if only subtly, from its self; and evolution may also favor a self-replicator which takes short-cuts including, at least theoretically, those which alleviate it of itself doing the replicating. In short, these two characteristics are dialectically entangled: self-replication puts evolution in play, and evolution puts self-replication in play.

This dialectical entanglement problematizes the third aspect of Pace's definition. Life is a system. A system implies something enclosed and whole. The question with regard to life would be, "What can count as the whole?" The question is familiar to both biology and the philosophy of biology; a provisional answer that identifies the whole with the individual is subject, on the one hand, to a whittling which

[3] Of course, this implies the necessity of interrogating what "genetic" means, a problem which for the moment I shall skirt.

pares the whole down all the way to the gene, and on the other hand, to an aggrandizement which expands it from populations all the way to the biosphere.[4] In the context of this chapter, more helpful than asserting a solution by fiat is to investigate the necessity of so wide a range of views; and this, I will try to show, follows very quickly from the interplay of the self-replicating and the evolving. As a consequence, *all* borders which would, in absolute fashion, define biological entities, from gene, virus, cell and organism to species, ecosystem, biosphere and life itself, are not only unstable but ultimately indeterminate. Such a recognition not only challenges the project of definition in this case; it also can only invite a reconsideration of the "is" in the instituting question, "What is life?" – even if my most serious attempt at that reconsideration is necessarily deferred, at least until the chapter on Darwin.

Finally, a fourth aspect is implicit to Pace's definition. He did not, in fact, require all living beings to be self-replicating and evolving; he required that they have the *capacity* to self-replicate and evolve. At any given moment, a living being may

---

[4] For example, we might consider the mite *Acarophenax tribolii*, the males of which impregnate their sisters in the womb of their mothers and die before being born. Can these males be distinguished from something like an organ? Likewise, Darwin's discussion of neuter insects in On the Origin of Species leads him to recognize selection as operating at the level of families as well as individuals (Chapter VII, p. 237 in the first facsimile edition), a circumstance he reconciles via relatedness. Yet earlier in the chapter (pp. 219 – 224) he had described slave-making ants absolutely dependent on slave ants of a different species – and might we not reasonably wonder, despite the absence of close relatedness, what here constitutes the whole? More generally, as I shall discuss below, Darwin inscribed this problem everywhere in his text as soon as he emphasized the dependence of one being on another in his understanding of the term "struggle for existence."

not happen to be doing either; and Pace would readily admit this. In framing life in terms of capacity or potential, however, we enter fraught territory for at least two reasons. First, potential ultimately eludes both observation and experiment – that is, it eludes the investigative tools of biology. Even if, according to an interpretation of a genome sequence, an organism has the capacity to do a particular thing, the fact that it is never observed to do so eventually will call into question the proposed interpretation; just as it would were the organism observed to do something for which, on our interpretation, it lacked the genetic potential. Concisely put, we cannot empirically or experimentally distinguish an absolute from a circumstantial (in)ability. Since the combinatorial scope of circumstances is indefinite, potential is contextually un-plumbable – all the more so when its temporal dimensions are considered. (How long does a biologist wait before deciding an organism can't do something in a particular context?[5]) Here the productivity of the discipline of biology should strike us: it is an accomplishment to be wondered at, when life is characterized in a way that reveals, from the very start, our methodology's necessary inadequacy.

The second reason this territory is fraught is related: should we not pause over a *being* defined in terms of a capacity or potential, an "is" that means "may be"?[6] As with the third

[5] On this score, an instructive example are the cryptoendolithic microbial communities in the Antarctic (e.g. Friedmann and Ocampo 1976), which may (detectably!) metabolize for only a few hours a year.

[6] I do not deny that being may be thinkable in these terms, a thinking which, for example, Heidegger invites in Being and Time. I am only suggesting that such a formulation necessarily thematizes the question of being, a question biology does not understand itself to be asking.

aspect of Pace's definition, life's systematicity, here again the broaching of this problem would require us to reconsider an "is" which is neither existential nor copulative, though mistakable for both; and as before, in this context, I shall defer to the Darwin chapter my most sustained approach to that formidable little word.

Instead, the goal of this chapter is, first, to demonstrate how biology itself calls into question what at the same time it takes as given, and second, to work out some of the consequences. In particular, I intend to examine actual biological examples of the abstract claims and arguments I have thus far made. To begin I will analyze a classic experiment in molecular biology: the work of Sol Spiegel-man's laboratory with the nucleic acid genome of the virus Qβ. This experiment will provide a concrete example of the ambiguities that necessarily surround the concepts of self and self-replication in biology. It will also demonstrate the tension between the self-replicating and the evolving, while showing that this tension necessarily puts into play familiar categories. The chapter will then track that play in a variety of biological projects, paying particular attention to the search for the minimal gene-set – genes often considered "essential" for life. This search will serve as a model by which to critique a particular vein of essentialism that predominates in modern biology, as well as the whatness, wholeness and systematicity of life, all three of which are asserted only at the considerable expense of a determinism that is secretly teleological and/or eschatological. Their assertion also becomes the unwitting caricature of a common understanding of our human selves, particularly in terms of individuality, identity and autonomy. The question then will be, can we think life – and ourselves – differently? Or, perhaps, are we despite ourselves already thinking them differently? If we are, what has driven us to do so?

II. Monstrous life

"'Exploitation' is not part of a decadent or imperfect, primitive society: it is part of the fundamental nature *of living things, as its fundamental organic function; it is a consequence of the true will to power, which is simply the will to life."*

Nietzsche, Beyond Good and Evil, aphorism 259

In 1967, Solomon Spiegelman and co-workers undertook, in the words of their title, "an extracellular Darwinian experiment with a self-duplicating nucleic acid molecule," the latter being, at least initially, single-stranded RNA that served as the genome of the bacterial virus Qβ. Having previously isolated the enzyme that replicated the viral genome (the "replicase"),[7] they placed the naked RNA of the virus in a test tube to serve as initial template; and along with it they added replicase enzyme and abundant, radioactively labelled ribonucleotide triphosphates (NTPs, subunits from which RNA is made) and commanded the RNA molecules to follow, in their

---

[7] What in this context does it mean to "replicate" a (viral) genome? It means that the replicase enzyme used the RNA that (at least formerly) served as the genome of Qβ as a template and catalyzed the polymerization of individual nucleotides (subunits of RNA) to one another in accordance with that template. The result is, in some way, a repetition of the template – though that repetition may not, it turns out, be an exact copy. We should ask, though notably the experimenters did not, whether the result is still a replica and the enzyme still a replicase.

words, "the Biblical injunction, multiply, with the biological proviso that they do so as rapidly as possible" (Mills, Peterson and Spiegelman 1967). The biological proviso was enforced by way of a series of transfers, 75 in total, in which a small aliquot from one test tube was used to seed a new test tube containing only replicase and NTPs (no RNA; therefore the template was provided only in the transfer). To further favor faster replication, the interval between transfers was decreased over the course of the experiment from 20 minutes to as little as 5. "The outcome is what might have been expected on *a priori* grounds," the authors wrote. By tracking the radioactive fraction, they showed that the viral RNA got progressively smaller, culminating in a molecule ~ 1/5th its initial size. By getting smaller, the individual RNA molecules could be replicated faster.[8] This was possible in part because, in a world replete with NTPs and environmentally available replicases, no selective pressure existed to compel the retention of actual genes. Indeed, analogous later work (with different viral RNA) reduced the RNA size even further and showed that the preserved stretches of sequence were not necessarily identifiable as genes at all but instead contained sites required for the binding of the replicating enzyme (Oehlenschläger and Eigen 1997).

First, observe the echo of Pace's 2001 definition of life as a self-replicating, evolving system in the title of Mills et al.'s paper ("An extracellular Darwinian experiment with a self-duplicating nucleic acid molecule"). This molecule, in the

---

8 The situation is akin to an exercise of copying a single sentence repeatedly. If a 20-word sentence should, through a series of "mistakes," be reduced to a shorter sentence, it will accordingly "replicate" faster, all else being equal.

context of a "Darwinian experiment," evolves; and it replicates, it makes a duplicate of itself. But in that phrase, is the "itself" a subjective or objective genitive? That is, is the self the subject that does the copying or the object that gets copied? The answer, it turns out, is complicated. It's "both," but with an important qualification: the subjective "itself" is not the same as the objective "itself." Thus, the self that copies is not only the nucleic acid molecule; it's the world of that molecule, a world which includes – but this list is *necessarily* not exhaustive – building blocks (NTPs), builders (the replicase enzyme), a communication-enabling medium (the aqueous solution), provisional boundaries (the test-tube walls), and occasional god-like intercessors (the scientists), among whose unstated duties is, for example, the maintenance of the laboratory at a temperature conducive to the operation of the enzyme.[9] Yet the self that gets copied is simply, on the authors' view, the RNA molecule.

This equivocation in the understanding of a self-replicating (or -duplicating) entity is pervasive. For example, autotrophs (i.e., self [auto] – nourishers [troph], like photo-synthetic plants or chemosynthetic bacteria) are often extolled, going back at least to Pirie (1937), as independent and autonomous, making them candidates for the earliest life forms on Earth, for terraforming other planets, or for survivors

[9] Note that maintaining the laboratory at such a temperature in turn implies, for example, electricity; and electricity implies a power plant; and a power plant implies, eventually, a universe. What Mills et al. describe as "the comparative simplicity of the system" is an illusion, one parodied by their own reference to "the universe provided to [the RNA molecules] in the test tube." In an important sense, the test tube does represent an entire universe – as its product and even, in a sense, culmination.

who by themselves could repopulate the world after a planetary-scale catastrophe. But as with the Qβ experiment, the self (*auto*) which gets nourished (the individual plant or bacterium, say) is subtly different from the self which nourishes. Though it generates energy directly from the sun or from a chemical potential difference, an autotroph still requires a world to supply it with the necessary materials and conditions; an autotroph's autonomy is an illusion if understood to suggest an independence from context. Indeed, it probably requires a *biotic* world to recycle and supply materials at necessary rates,[10] in order to stave off – as deferral, not absolute deliverance – an eventual life-ending equilibrium that it by itself cannot jar.[11]

But what does "by itself" really mean? If it's self-replicating, it's not by itself – it's surrounded by so-called replicas. Now, however, a second important equivocation in the use of the word "self" emerges: the self, now understood as object, as that which is copied, proves not to be necessarily self-identical. At least, it's not self-identical if it's also evolving.

---

10 The necessity of these rates may be for reasons other than purely physical or chemical ones (if such pure reasons ever really exist); it may reflect coevolutionary constraints, e.g., the circumstance of present-day autotrophs having adapted to present-day *bio*geochemical rates and processes.

11 This contention is in fact not radical, even if it is widely overlooked. In his 1968 study of biological thermodynamics, Harold Morowitz made a similar argument as one of his basic generalizations about life: "Generalization VII: Sustained life under present-day conditions is a property of an ecological system rather than a single organism or species." He continued: "A system of primary producers [= autotrophs] would grind to a halt from CO_2 exhaustion unless autolysis produced CO_2 at a sufficient rate. This does not appear to occur."

This is clearly the case with the Qβ RNA: at the end of the experiment, the RNA is manifestly different than it was at the start of the experiment. The self that got replicated got replicated with a difference. Herein lies the tension between the self-replicating and the evolving: the perfect self-replicator – if such is even truly thinkable, if the mere act of replication doesn't itself constitute a transformative difference no matter its exactness – cannot evolve. The autotroph left alone on a devastated planet, if it cannot evolve, will eventually exhaust its resources and go extinct; whereas, if it can evolve, it may stave off that extinction – again as deferral, not absolute deliverance – albeit at the price of changing, perhaps into something dramatically different, something that's no longer even an autotroph. Indefinite deferral (self-replication) requires difference (evolution).

This two-fold equivocation in "self" and the taut relationship between the "self-replicating" and the "evolving," between deferral and difference, together reveal, not so much the inadequacy of Pace's working definition, as the strange nature of its fruitfulness.[12] For a moment, the paradox embedded in Pace's definition allows us to touch what at that very moment also escapes our grasp – life as something both recognizable and protean. The experiments of Mills et al. accomplish something similar: they demonstrate – though this

[12] I cannot help but observe that the 'entity' – if that's the right word – we're chasing in order to define life has the subtlety, the pleomorphism and the elusiveness of the sophist literally hunted, albeit unsatisfactorily if not unsuccessfully, by the Eleatic stranger in Plato's *The Sophist*. There the question of what a sophist is requires an investigation of Being and Non-being. Here the question of what life is may require something similar – though I have exiled that inquiry to the chapter on Darwin.

is almost entirely overlooked or forgotten – how malleable our categories are when under evolutionary pressure. Thus, in Spiegelman's laboratory typical criteria by which we might understand a virus, such as infectivity (for which Mills et al. tested) or the presence of genes, were lost. As a result, we might wonder what exactly the resultant entity *is*. Can we still call it a virus, though apparently incapable of infection? Does the surviving strip of nucleic acids still count as a genome, though no longer encoding genes?[13]

This problem is allegorized by the colorful sobriquet that came to be given to Qβ's eccentric spawn: "Spiegelman's monster." That nickname implicitly acknowledges a problem that it also, in the same gesture, slips past. On the one hand, the nickname allows us to avoid formally naming a being which we no longer know how to categorize. We neither call it a virus nor admit that it isn't one. On the other hand, in apparent playfulness, the nickname also supplies the searched-for category that it admits can't be found. But the supplied category is a special one reserved for the exceptional, the fabulous and, indeed, for the uncategorizable. It is the category of monster. Not coincidentally, the identification of it as a monster is tied to switching its parentage: this progeny has ceased to be Qβ's and has become man's (in fact Spiegelman's).[14] A demand for classification, which since

---

[13] Indeed, we ultimately must ask whether the term "gene" is meaningful *at any moment* in the context of Mills et al.'s experiment. Do nucleic acids encode genes in an environment that precludes their expression? Are genes truly context-independent categories?

[14] One should not overlook the implications of this switch for self-replication: that which does the replicating is no longer clearly the RNA, if the monster is Spiegelman's.

Darwin has explicitly been understood in biology as the pursuit of a "natural system," is thus simultaneously met and not met: a "category" has been supplied, albeit only half-seriously; and that light-heartedness is tolerable because the entity in question, an artifact produced by man, is not natural and needn't after all be included in a natural system.[15] Yet, this solution – which is all the more insidious for its seeming insouciance – has successfully covered up what is most interesting about this experiment: *that entities in which so-called self-replication (deferral) and evolution (difference) are paired end up transgressing the definitional boundaries by which we know them*. The nickname thus veils its own necessity. We fail to ask why a regular name won't do.[16] Instead we precisely *don't* linger over the problem, not only of what Spiegelman's monster *is*, but also of how it came to be that. A half-century can then pass without anyone dwelling on the significance of a being which evades the categories of our thought by "self"-replicating and evolving.

A second detail of this experiment also requires comment and interpretation. "Spiegelman's monster" did not jettison everything. Indeed, as later experiments showed, under the circumstances of the experiment it was compelled to retain specific nucleotide sequences whose presence was required in order for the replicase enzyme to recognize, bind and copy the

---

[15] Such a reading of course takes for granted the human exclusion from nature – a gesture the dangers and problematic assumptions of which I have already sketched in the Parodos and address most directly in the third chapter on ecological restoration.

[16] Much less do we ask whether a formal name in the end is truly distinguishable from an informal nickname. An upshot of this chapter will be the suggestion that it is not.

RNA. Note that this is not some general sequence of universal applicability; the "recognition sequences," as they're called, had value only in relation to the particular replicase enzyme with which they coevolved. A different enzyme would have required different sequences. As a consequence, the strips of RNA preserved by and as Spiegelman's monster were the result neither of mere chance nor of exclusively physical or chemical criteria "in the abstract." There *were* physical or chemical criteria, but their import was *historical*: they reflected the coevolution of the virally-encoded replicase with the viral nucleotide sequence.

How does this work? In order to be "self-replicating," a biologist would point out, it is necessary to distinguish your "self" from others. In the context of a virus-infected cell, this can be tricky, since considerable quantities of cellular RNA are available that are (at least ostensibly) not viral. Hence strong selective pressure is exerted on viral enzymes to replicate only nucleotide sequences which have specific motifs, motifs found in their own RNA but rarely in cellular RNA. Thereby can the viral enzyme selectively replicate viral rather than (or at least in preference to) cellular RNA, in accordance with at least a naïve understanding of "self-replication."

That the conditions of the experiment by Mills et al. preempted any need for such discernment – since there was *only* viral RNA to replicate[17] – was irrelevant; the prior

[17] But was there? Couldn't we read it otherwise, that the experiment reveals an enzyme which could always be yet more discerning of its "true self"? Or might we instead argue that it is more discerning of its "true self" than we are, even when it replicates what we call "cellular RNA"? In short, when we

coevolution of replicase and sequence operated as a historical constraint on the experiment's possible outcomes. In this particular context, there actually *was* a limit to how small the monster could become and to what the resultant nucleotide sequence could be: recognition sequences had to be retained or no replication would take place. This limit, however, was never an absolute limit for any entity or for any set of conditions; it pertained only for this entity with its peculiar evolutionary history and under these particular conditions. A similar experiment with a different replicase enzyme, requiring different recognition sequences, would give broadly comparable results (i.e. favoring RNA molecules that became smaller) while nonetheless selecting for different specific sequences and sizes – as Oehlenschläger and Eigen (1997) confirm.

In summary, then, Spiegelman's monster represents two important problems for us. First, an evolutionary process "culminates" – albeit always provisionally – in an entity for which we have no clear category. Second, this process is not simply reducible to physics and chemistry; a biotic history of coevolution is implicit to it. Both of these problems fall out of the equivocal meaning of "self" in "self-replicating," insofar as the objective self, the one that's replicated, changes in response to a subjective (replicating) self whose boundaries are far from determinate. Yet, in the extensive literature that Spiegelman's monster has generated, neither problem has been thematized or, for the most part, even noticed. This is not because biology is unfamiliar with them, however. Far from it: modern biology

---

designate RNA as "viral" or "cellular," have we not again fallen into a sort of essentialism which this experiment invites us instead to critique?

largely rests on a theory inspired by versions of precisely these two problems, which correspond closely to the two great insights which struck Charles Darwin in the Galápagos and led him to the theory of natural selection.

What were Darwin's two great insights in the Galápagos? First, Darwin recognized that he couldn't define species – that it defied clear categorization: "Many years ago, when comparing, and seeing others compare, the birds from the separate islands of the Galápagos archipelago, both one with another, and with those from the American mainland, I was much struck how entirely vague and arbitrary is the distinction between species and varieties" (On the Origin of Species, Chapter II, p. 48). This led him, in the last chapter (XIV, p. 485), ultimately to advocate in strident and daring terms that we abandon "the vain search for the undiscovered and undiscoverable essence of the term species."[18] Darwin would certainly be surprised by how little heed biology has given this advice, even while it has ostensibly embraced the theory that motivated it. Instead, the century-and-a-half that has elapsed since the publication of The Origin has witnessed repeated, widely-embraced attempts to formalize the species concept once and for all, whether based on reproductive compatibility and isolation (which Darwin himself had considered and wisely rejected on the grounds that they were always matters of degree) or on an arbitrary amount of sequence dissimilarity in

[18] "In short, we shall have to treat species in the same manner as those naturalists treat genera, who admit that genera are merely artificial combinations made for convenience. This may not be a cheering prospect; but we shall at last be freed from the vain search for the undiscovered and undiscoverable essence of the term species."

particular genes (e.g., a 3% difference in ribosomal RNA genes).[19] No definition has succeeded, yet biology keeps being drawn to flitter around this particular flame – and it behooves us to wonder why.

The second, equally great insight of Darwin in the Galápagos was that living beings were typically more important than physical conditions in shaping each other; that is, he emphasized the significance of a biotic history of coevolution:

"... [F]or it may be asked, how has it happened in the several islands situated within sight of each other, having the same geological nature, the same height, climate, &c., that many of the immigrants should have been differently modified, though only in a small degree. This long appeared to me a great difficulty: but it arises in chief part from the deeply-seated error of considering the physical conditions of a country as the most important for its inhabitants: whereas it cannot, I think, be disputed that the nature of the other inhabitants with which each has to compete, is at least as important, and generally a far more important element of success." (Chapter XII, p. 400)

Characteristically, this insight, too, is either forgotten or its deep implications pass unnoticed. Thus, it is common for

[19] This latter fell apart when a single organism, *Haloarcula marismortui*, was discovered to have 16S ribosomal RNA genes which differed by > 5% (Amann et al. 2000).

organismal traits to be interpreted as self-evidently adaptive[20] to a particular physical condition, as if that condition were encountered alone in its abstract purity. Under what extraordinary circumstances, however, can we be sure that a trait understood by us in terms of a particular physical condition, such as extreme cold, is now and was historically in fact an adaptation to that condition, independent of all the other conditions, biotic and abiotic, experienced by and contextualizing the ancestral organisms?[21] A similar assumption underlies research into abiogenesis, always understood as the emergence of life from nonlife. Just as serious as the question of why biology continues to join the search for the essence of the term "species" is the question of why it would be attractive to biology to reduce organisms to abstract and lifeless contexts, whether in the understanding of a trait or in pursuit of life's origin.

Both of these questions are central not only to this chapter but to the book as a whole, and so the deepest answer will

---

[20] The language of "adaptation" is already problematic, as Gould and Lewontin (1979) have pointed out.

[21] I choose the example of the cold precisely because here Darwin breaks with his own thought: "When we reach the Arctic regions, or snow-capped summits, or absolute deserts, the struggle for life is almost exclusively with the elements," he writes in Chapter III (p. 69). He reiterates this conclusion more stridently at the end of the chapter (p. 78): "Not until we reach the extreme confines of life in the arctic regions or on the borders of an utter desert, will competition cease." Based on experimental field work during Arctic winter, as well as additional laboratory work, Wells (2006) argued that Darwin was mistaken and that even under those (for us) extreme conditions the struggle for life is always also a struggle with life, in both the antagonistic and collaborative senses of the preposition "with".

require a prolonged and subtle unfolding.[22] As a beginning towards that, however, it is important to realize that the questions are inter-related, as also were Darwin's two insights in the Galápagos. For those insights are but sides of a coin. Species elude abstract and enduring definition because the beings that constitute them are ever shaping and being shaped by each other. If this is always, strictly speaking, an accidental feature, it is nonetheless for life always a necessary accident. It is the corollary of an evolutionary process which renders provisional every taxonomic category from "species" to "life" itself, since these categories are always in play to the precise degree that evolution remains possible. It is, in other words, the already-mentioned equivocation of "self" writ large. Darwin's plea in the final chapter of The Origin that we abandon the search for the essence of the term species is accordingly a *necessary* conclusion to his argument: taxonomic categories are indeterminate because of evolution, the biotic context of which, Darwin asserted on more than one occasion, is at least as important and usually far more important than the physical[23] – assuming, *concesso non dato*, the two contexts can be separated absolutely in the first place. Ignoring these facts,

[22] The Cartesian dimensions of these questions will be explored especially in Chapter 3, while their even deeper Aristotelian roots will be discussed in Chapter 4.

[23] For example, in Chapter XI, p. 350: "The degree of dissimilarity [of inhabitants of different regions] will depend on the migration of the more dominant forms of life from one region into another having been effected with more or less ease, at periods more or less remote; – on the nature and number of the former immigrants; – and on their action and reaction, in their mutual struggles for life; – the relation of organism to organism being, as I have already often remarked, the most important of all relations."

or extrapolating away from them, is fundamentally to misconstrue life – and thereby to risk substituting a rigid parochialism, in which life is understood ultimately as an isolatable thing, for what is instead processual and relational and, accordingly, protean. It is to mistake as monstrous what is paradigmatic and to dismiss the accidents of context, whether present or historical, as inessential and therefore inconsequential.

Let me, however, take a step back and actually establish what is paradigmatic about Spiegelman's monster, lest it appear that I am endeavoring to demonstrate too much from a mere monster. A biologist might dismiss my argument on two grounds, both in some ways implicit to the substitution of a nickname for a formal name. First, the biologist might dismiss these problems as pertaining, after all, only to a non-living entity – a stretch of nucleotides itself derived from a mere virus. Such a dismissal risks being hoist by its own petard, however, since to admit the existence of non-living but still evolving and "self-replicating" entities is itself to blur boundaries between traditional categories[24] (as the discovery of Pandoraviruses – viruses with traits and a complexity that approach cellular forms – has recently also suggested; Philippe et al. 2013). If instead it is objected that the monster is, after all, Spiegelman's – i.e., that it is possible only in a laboratory and hence is irrelevant to the study of the natural world and to a critique of universalizing characteristics of life there applied – then a multitude of phenomena, not sequestered in

---

[24] Such an objection dissolves Pace's operational definition of life, for example.

laboratories but strikingly similar to Spiegelman's monster, must be accounted for.[25]

The closest examples – the examples closest both to Spiegelman's monster and to us – are the abundant human endogenous retroviruses (HERV), approximately 98,000 of which have squatted 5 – 10% of the human genome. As typically conceptualized, HERV consist of three genes (*gag*, *env* and *pol*) flanked by long terminal repeats (LTRs).[26] They are replicated in one of three (not necessarily mutually exclusive) ways. In the first, reinfection, all three genes are expressed[27] to synthesize a viable virus particle capable of newly infecting a cell. The second, retrotransposition in cis, can be thought of as a type of parasitism in which only two of the three encoded genes are required, the gene product of the other (*env*) being supplied environmentally, generally by way of a second, proximate virus. Yet more extreme is the third option, retrotransposition in trans, in which none of the viral genes are required, all necessary proteins being supplied environmentally.

Among recognized HERV families, elements of the HERV-K (HML2) family have both a low tolerance of nonsynonymous substitutions[28] in the *env* gene (the ratio of

---

[25] I again set aside, until the third chapter, a critique of the assumption underlying such an objection, the firewall dividing man from nature.

[26] LTRs are stretches of repetitive nucleotide sequence without expected characteristics identifying them as encoding genes; they are commonly associated with the insertion sites of viruses into the genomes of their hosts.

[27] By "expressed" I mean "translated into protein."

[28] A nonsynonymous substitution is one in which a different amino acid is encoded as a result of a genetic change. Because the genetic code is highly redundant – on average, there are ~ 3 codons for every amino acid – a genetic

nonsynonymous to synonymous substitutions [d_N/d_S] is significantly less than one, indicative of purifying selection) and few examples of acquired, disruptive stop codons in any of the three genes, suggesting that these genes largely remain active and hence that the elements are replicated predominantly by reinfection (Belshaw et al. 2004). On the other hand, elements of the HERV-L family have lost their *env* gene while members of the HERV-K (HML3) family have a high d_N/d_S ratio for *env* but a low one for *pol*, consistent with retrotransposition in cis as the dominant replicative mode in both families (Belshaw et al. 2005). Similarly, ~ 90% of the elements in the HERV-H family have major deletions in *pol* and *env* (and a majority also has deletions in *gag*; see Supplementary Material in Belshaw et al. 2005), which implies replication is limited to retro-transposition in trans.

Notably, of 17 HERV families examined, the four with greater than 200 copies in the genome appear to have used retrotransposition in cis or trans as their predominant or exclusive mechanism of replication; of these, HERV-H is by far the best represented, with 1306 copies (compared to 418 or less for the other three families; Belshaw et al. 2005). Moreover, the analyses are biased against detection of HERV replicating by retrotransposition in trans, since the HERV are identified by phylogenetic comparison of conserved regions in the genes – retention of which may be selectively neutral for retrotransposition in trans. To consider an extreme case, if an element replicated exclusively in trans, it is possible that no selective cost would accompany the loss of all encoded genes,

change often does not correspond to a change in the amino acid sequence of a protein; in those cases, the substitutions are synonymous.

provided that an intact promoter and structural motifs for expression and packaging of RNA were maintained. With flanking LTRs at both ends, such minimization might be accomplished by recombinatorial deletion. Solo LTRs, 10-100 times more abundant in the genome than HERV, may thus represent not degenerate or inactivated forms (as commonly interpreted), but rather a currently successful, if monstrous, evolutionary trajectory.

But is it in fact monstrous? By exploiting environmentally-available materials, even if biotic ones, both Qβ and members of the HERV-H family could dispense with genetic information rendered redundant by their environment. So long as such loss is not selectively disadvantageous, it does not have to be selectively advantageous (though it can be, e.g., by enabling faster replication, as in the case of Qβ). The bestiary of parasites, symbionts, pathogens and societies is littered with analogous examples, usually classified under the (somewhat misleading) rubric of "reductive" evolution. To choose just a few dramatic cases: among the Bacteria, the (apparently) obligately intracellular leprosy bacillus *Mycobacterium leprae*, in addition to missing many common metabolic capacities, also has almost as many pseudogenes[29] (1116) as protein-coding genes (1604; Cole et al. 2001); among the Archaea, *Nanoarchaeum equitans*, an (apparently) obligate parasite or

---

[29] Pseudogenes are genes which, for a variety of reasons, are thought not to be expressed in the cell and hence are considered "defective." They are (in parallel) often interpreted as "functionless," though such an interpretation may too narrowly conceive of what their function could be. (Scrap paper may be functionless with respect to its prior use, but quite functional as scrap.)

symbiont of the crenarchaeon *Ignicoccus*, is missing most or all of the known genes needed for *de novo* biosynthesis of lipids, cofactors, nucleotides and amino acids, not to mention many of those involved in core cellular metabolism, including glycolysis, the TCA cycle and the pentose phosphate pathway (Waters et al. 2003); and among the Eukarya, the parasitic microsporidian *Encephalitozoon cuniculi* is also missing (like *N. equitans*) most of the known genes concerned with amino acid, nucleotide and fatty acid biosynthesis and many involved in the TCA cycle (Katinka et al. 2001).[30] Yet the conservation of these genes and pathways in closely-related organisms (or as pseudogenes) suggests that most or all of the missing genes were formerly present in an ancestral organism, implying gene loss. Presumably the missing genes were dispensable because the host – a biotic environment *par excellence* – supplied their functions, as with Qβ and HERV-H elements.

My point is that exploitation, no matter that from our perspective it may appear short-sighted, is not the evolutionary exception, but in many ways the norm. That to which an organic being adapts of necessity and in all cases includes other organic beings. If Qβ's adaptation to an environment where nucleotides and replicases were replete seems artifactual, is it fundamentally different from the examples given above, or for that matter from the adaptation of heterotrophs to an environment suffused with biologically fixed carbon? In the latter case, it might be argued that, unlike Qβ, heterotrophic dependence on the activities of photosynthetic (or chemo-

30 These examples were chosen because life is commonly (and always problematically) organized into three domains, the Bacteria, Archaea and Eukarya; hence I have given an example from each domain.

synthetic) organisms is not a derived trait, heterotrophs never having encoded genes subsequently displaced by environmentally available gene products. But, of course, some did, including trypanosomatid parasites, which once contained plastids[31] but lost them (Hannaert et al. 2003), apicomplexan parasites with the plastid-derived hohlzylinder, and heterotrophic plants such as *Epifagus virginiana*, the plastids of which no longer encode photosynthetic genes.

As implied by some of these examples, the phenomenon of acquired reliance on a biotic environment is not limited to the microbial world: among metazoa, slave-making ants described by Darwin are a dramatic example of one species exploiting another to the point of utter and even disastrous dependence upon them, not unlike the accrued dependence of Qβ's descendants on Spiegelman's lab:

"This remarkable instinct [slave-making] was first discovered in the *Formica* (Polyerges) *rufescens* by Pierre Huber, a better observer even than his celebrated father. This ant is absolutely dependent on its slaves; without their aid, the species would certainly become extinct in a single year. The males and fertile females do no work. The workers or sterile females, though most energetic and courageous in capturing slaves, do no other work. They are incapable of making their own nests, or of feeding their own larvae. When the old nest is found inconvenient, and they have to migrate, it is the slaves which determine the migration, and actually carry their masters in

---

[31] Plastids are the organelles typically associated with photosynthesis in a eukaryotic cell. Since plastids normally have an enclosed genome, their loss or reduction implies gene loss.

their jaws. So utterly helpless are the masters, that when Huber shut up thirty of them without a slave, but with plenty of the food which they like best, and with their larvae and pupae to stimulate them to work, they did nothing; they could not even feed themselves, and many of them perished of hunger. Huber then introduced a single slave (*F. fusca*), and she instantly set to work, fed and saved the survivors; made some cells and tended the larvae, and put all to rights." (Chapter VII, p. 219)

One could add to this many other examples of extreme, selectively-favored loss of function, even going so far as (provocatively) to invite the inclusion of worker sterility among social insects or the differentiated somatic cells of metazoa.[32] In each case, a loss of function has been accommodated by way of a supplement (or, in the parlance of genetics, a complement) supplied environmentally, albeit with the environment necessarily understood as biotic. Seen in this light, dependency is not an imperfect or decadent corollary to

---

32 Another example which Darwin gives are the "complemental males" found among some cirripedes (barnacles), "for the male is a mere sack, which lives for a short time, and is destitute of mouth, stomach or other organ of importance, excepting for reproduction" (Chapter XIII, p. 441). In his Monograph of the Sub-Class Cirripedia he went further: "These males... consist of a mere bag, lined by a few muscles, enclosing an eye, and attached at the lower end by the pupal antennae, it has an orifice at its upper end, and within it there lies coiled up, like a great worm, the probosciformed penis... there is no mouth, no stomach, no thorax, no abdomen, and no appendages or limbs of any kind... I know of no other animal in the animal kingdom with such an amount of abortion" (as quoted by Stott 2003, p. 220). Here, even naming can become a problem: while "cirripede" means "curled foot," Darwin discovered one which was "naked, apodal [footless], with a suctorial mouth & parasitic in a double way within another cirripede" (*Ibid.*, p. 109).

exploitation, but rather a characteristic of exploitation and, if the continuing allusion to Nietzsche can be forgiven, of the will to life. But it is in part precisely this adaptive significance of exploitation and of dependency that blurs distinctions between categories, whether of species, organisms or even individuals.[33] It is here that we see clearly the convergence of Darwin's two great Galápagos insights, on the one hand, and on the other, a dialectic between replicating and replicated selves which calls into question the self's proper identity.[34]

If we do not know precisely what Spiegelman's monster is, or what a species is, or what a HERV is, or a LTR, or any of the other examples of fabulous creatures listed in this barely-begun bestiary, up to and including life itself, then we must recognize that such ignorance is not merely a bothersome and temporary inconvenience. It is rather a positive fact about life. Life is shaped not only by some endogenous potency like its genetic code;[35] it is equally shaped by its (always already biotic) context. If it endures, it endures by pairing self-replication with evolution, inheritance with modification and deferral with difference. As a result, however, we do not know what it is.[36]

---

[33] It is equally this adaptive significance that renders the desired distinction between physical and biotic environments suspect.

[34] That is, in terms of the already-discussed equivocation in the understanding of "self," what we are describing is precisely their dialectical entanglement: the deficiencies of an objective "self," the self that gets replicated, are compensated for by the subjective but indeterminately bounded replicating "self" which enables those deficiencies in the first place.

[35] Contra the claims of, e.g., Michael Archer, the species revivalist discussed in the Parodos.

[36] It might be objected that so-called living fossils prove that not everything is always changing. But this is a more complicated claim than it appears, as

We do not even know that it is a *what*. We cannot draw decisive, impermeable limits around it any more than we can around species. Instead, whether with Spiegelman's monster or with slave-making ants, we are on the cusp of glimpsing other possibilities. For an instant, a protean figure squirms at the very tips of our fingers, if only we could see and bring it into focus; but what it is, is blurred, and to focus it is to misapprehend it. To call attention to it, to name it, is to utter the word which breaks the charm by which our quarry had, almost for an instant, been mesmerized; and with that same word it is to risk mesmerizing ourselves.

Darwin acknowledges: "I have found it difficult, when looking at any two species, to avoid picturing to myself, forms directly intermediate between them. But this is a wholly false view; we should always look for forms intermediate between each species and a common but unknown progenitor" (Chapter IX, p. 280; see also Chapter XIV, p. 464). To illustrate this claim, however, he bizarrely gives the example of the rock pigeon, which he takes to be both common progenitor and descendant: "...for the parent rock-pigeon now lives," Darwin writes in Chapter X (p. 335), seemingly unaware that he is using the word "parent" as well as the word "rock-pigeon" figuratively. The "parent" "rock-pigeon" would accordingly be analogous to the living fossil (a term which Darwin, incidentally, uses elsewhere; e.g., Chapter IV, p. 107 and Chapter XIV, p. 486, where "it will aid us in forming a picture of the ancient forms of life") though it is unclear whether it is so in fact or simply by decree. The problem with living fossils is precisely that the presumed ancestor is inaccessible except as *represented* by the living and by the fossils; and to rationalize that representation as founded not on similarity but on sameness can only be a rash gesture, one which confuses the picture, image or representation with its ostensible subject.

III. Are the essentials bare?

My argument so far has focused on the unavoidable tension between the "self" which is replicated and the "self" which replicates, one of the consequences of which is evolution and the pairing of deferral and difference. As a result, the phenomenon of life unavoidably evades determinative boundaries, in two senses: it evades categorical definition, whether as species, virus or even (ultimately) as living; and, relatedly, it evades an absolute distinction between self and other. The self that replicates is not the same as the self that gets replicated; and the boundaries of that first self are anything but fixed. Moreover, that first self, inescapably contextualized by a *biotic* environment into which it is thrown, necessarily inflects what that second self is and becomes, up to and including its transgression of familiar categories; and vice versa. So goes the dialectical dance. As much as biology has periodically acknowledged this dance – notably in Darwin's two key Galápagos insights and in the evolutionary theory which emerged in part from them – it has also been quick to overlook it, often by resorting to a nominalism whose provisional nature is soon forgotten. As a result, biology has repeatedly sunk back into the quicksand of essentialism, for example by repeatedly endeavoring to determine the essence of the term "species" or even by extrapolating a universal biochemistry – in the strong sense of "universal," i.e., a biochemistry which alien life would also evince – from characteristics of biochemistry currently found on earth (as Pace 2001 goes on to do). The very distinction between the disciplines of biology and ecology is already suggestive of this mistake, of the desire to imagine and understand life "in itself" and thus outside of all context.

The efforts to contain life within definitional borders follow immediately from its assumed whatness and wholeness. An unbounded whole is not a what, and an unbounded what is not a whole. Yet the fragmentation of the study of life into myriad disciplines and subdisciplines spanning vast scales – from individual molecules to an entire planet – hints at the obscurity of this what and whole. Accordingly, in a characteristic gesture not limited to the study of life, what cannot easily be found synoptically is instead sought diachronically, and life is reduced to imagined physical and chemical conditions that pertain not only to what it is but also to the context in which it is found, if not now, then primordially. The absolute distinction between the living and the non-living whose very elusiveness requires a disciplinary cascade from the biosphere to the gene is thereby posited as a necessary historical fact. That no living being has ever been, or ever will be, encountered outside of a context which includes other living beings is dismissed as an accidental, hence inessential, feature of life as we happen to know it now. But what if we were to consider instead that this is, in Nietzsche's words, the "original fact of all history"?[37] Or, equivalently, what if we refused to dismiss this fact as merely accidental? – not because it isn't accidental, but because accidents are, as it were, essential to understanding life. If we cling tenaciously to this fact of our experience, if we likewise cling to a healthy

[37] The quotation is again from Nietzsche, aphorism 259: "Assuming that this is innovative as theory – as reality it is the *original fact* of all history: let us at least be this honest with ourselves!"

suspicion of the desire to isolate every living thing from its brethren and ultimately from its context, we shall certainly see life in a new way, albeit one not easily reduced either to a what or a whole.

The alternative procedure, which prevails within all biological disciplines, instead neutralizes the contingency of the biotic environment by an arbitrary imposition of borders whose own accidental character passes unnoticed, or at least uncritiqued. Such an imposition confuses as the requirements of life what are in fact only the requirements of definition – requirements with their own history and context, particularly within the western philosophical tradition. But life, if understood as both self-replicating and evolving, may be ill-fitted to that tradition, a tradition whose subordination of the accidental to the essential mirrors its parallel subordination of the body to the soul and the finite to the eternal. In this tradition, what is clear and distinct – that is, what is well-bounded – signals at least the legitimacy, if not the truth, of a thought; and almost forgotten is the question whether such clear and distinct borders truly characterize not only the thinking thing but the extended, material and embodied thing – in short whether they characterize the living world. And not just a living world, but a living world which, by virtue of being living, is also evolving. Is it possible for us, however, simultaneously to insist on clear and distinct boundaries and on evolution? Is it not rather that the borders on which definition depends, evolution, at least potentially, transgresses? To require such borders nonetheless is to demand that evolution itself be contained within them. Such a solution can only deliver us to an untenable predicament: life, defined as an evolving system, can be defined only if it isn't an evolving system.

From within this quandary, it is no mere accident – i.e., an accident to be lightly dismissed and left uninterpreted – that 20th and 21st century biology has repeatedly posited both a beginning and an end to evolution.[38] For example, Carl Woese, one of the 20th century's greatest biologists and thinkers of evolution, was nonetheless compelled in his discussion of the evolution of ancient cells and of the origin of life to claim repeatedly that "modern cells are fully-evolved entities" (Woese 2000, p. 8395; see also Woese 2002, p. 8742). He never seems to have reflected on that compulsion or on what he could possibly mean by a "fully-evolved" entity. Taking for granted that life had an opening boundary, a beginning, he required a closing boundary, a single target at which ancient cells, as it were, aimed. He needed a telos (τελος) in order to complete the project whose beginning (archē, ἀρχη) he was at the same time postulating. Yet, it is a testament to the rigor of his thought that he couldn't help but problematize his own assumption by admitting that fully-evolved cells might nonetheless be subject to "relatively minor... variations on existing cellular themes." How the "fully-evolved" cell, thus qualified, could immure itself from that "principle, by which each slight variation, if useful, is preserved…" (The Origin, Chapter III, p. 61) he failed to ask.

---

[38] Nor is the familiar claim that human beings at last are in a position to control evolution unrelated, as alluded to in the Parodos and discussed in more detail in Chapters 2 and 3. This claim, however often repeated, has always proven premature, a fact which should (but does not) give pause to its proponents.

Such a view of life requires that a circumstance exist, whether presently or historically, in which life emerges on an inert environmental stage insulated from evolution by its exclusively chemical and physical character. That character is taken as prior to life, both historically and logically, with few wondering whence came these seemingly obvious categories of the physical and the chemical in the first place, much less whether their derivation wasn't by way of a prior conception of the living, if only through the impoverishment or negation of the animate into the inanimate. Instead, the inanimate is taken to provide the firm foundation that neither replicates nor evolves and on which animation would rest. Awkwardly, this view requires a conspiracy of accidents to trigger evolution and life into being – even though accidents thereafter will be conspicuously excluded from life's definition, for unexamined reasons ultimately philosophical in nature. Life so understood flashes into existence out of non-life – out of a category which, we always forget, is born with it. An absolute distinction that we make would retroactively establish the very stage on which the supposed emergence took place. But what if that stage does not exist and has never existed? What if life's constant adaptation to itself, its striving to be itself only by changing itself, its marriage of self-replication to evolution and deferral to difference, is so constitutive of the biological dynamic as to preclude a distinct beginning or end?[39] If we forsake the

[39] Notice that there are two directions in which one might read this possibility: that there has always been life, and that there has yet to be life. I appear to privilege the former reading in my insistence that the environment is always biotic – an insistence at least partially strategic in nature and aimed at countering the prevailing view of abiogenesis (life born out of non-life,

assumption (or myth) of a discrete origin of life, as of its conclusion or perfection, will we not also have forsaken, necessarily, a definition which, deterministically, *must* universalize particular aspects of life by paring it of its "essential" accidents? Will we not have forsaken as well that life is both a what and a whole?

But rather than asking such questions from some abstract and seemingly theoretical vantage, let me instead proceed as before, and show how such questions emerge out of biology's own practice, even if the discipline remains insightfully blind with respect to them. Just as earlier the disruptive tension of the self-replicating and evolving was examined particularly by way of Spiegelman's monster (as much allegory as example), so now can we consider how the problem which I am describing – the pursuit of an essence which the pursuit itself both supplies and supplants – appears in, and is covered up by,

---

out of an abiotic milieu). Greater rigor requires a somewhat different claim, one to which I will return at the end of this chapter: that the environment of "life" – necessarily in scare-quotes or, if one prefers, under erasure – is always already *almost* biotic. "Life" is always on the very threshold of punching through and becoming, as it were, life. But that threshold – or, better, horizon – is never crossed, even if it has yet to be wholly abandoned. On this reading, evolution describes the repetition, always with a difference, of "life" striving, and failing, to be itself – and the accomplishment of "life" is, at best, always the deferral of life's accomplishment. Inheritance, which is never simply familial or genetic, is an important manifestation of this difference and deferral. Indeed, if we think about this in terms of Derrida's neologism *différance*, which combines difference and deferral – or rather, admits their inseparability – we will recognize in this one word the very dance between evolution (difference) and self-replication (deferral) that is put forward as defining life even if, ironically, it is also what prevents life's definition. For more in this vein, see Chapter 2 on Darwin.

modern biology. For this purpose my example (and allegory) will be the persistent endeavor to identify the "minimal gene set that is necessary and sufficient for supporting cellular life," as Mushegian and Koonin (1996) characterized a first attempt since widely emulated (for recent reviews, see Luo et al. 2014 and Xavier et al. 2014). This quixotic quest, commonly tied to an archaeology of a primitive cell, is also explicitly associated with the identification of so-called essential genes (an association that Mushegian and Koonin themselves inaugurated).[40] As we shall see, however, the essentiality that emerges is of a curious and interpretable character. In particular, its logic, which would pare life of its accidents, succeeds only in whittling life down either to nothing or to a stopping point that is itself secretly accidental.

When Mushegian and Koonin initiated this project, their approach was to compare the whole genomes of two parasitic bacteria, *Mycoplasma genitalium* and *Haemophilus influenzae*, at the time among the smallest microbial genomes known. The minimal gene set, they reasoned, would at least in approximation be those genes found in common between the two organisms. Yet Mushegian and Koonin made an interesting exception: they eliminated from consideration

---

[40] For example, they wrote in the abstract, "…[T]he genes that are conserved in these two bacteria are almost certainly essential for cellular function. It is this category of genes that is most likely to approximate the minimal gene set." Notice that here essential genes would delineate a minimal gene set. A few years later, Mushegian (1999) argued the converse: "Reconstruction of a minimal protein set uses the knowledge of existing genomes to approximate the essential set of survival genes." The extensive literature that this field has spawned itself appears to use "minimal gene set" and "essential genes" interchangeably.

specifically the genes which, though common to the two bacteria, were peculiar to the organisms' parasitic lifestyles. Their reasoning was basically that, since not all life is parasitic – *but is this true?*[41] – such genes could not be essential. Their minimal gene set thus began from the assumption that the cell's specific adaptations to its environment (in this case, each parasite's adaptations to its host, to its *biotic* environment) are not essential to it. Yet, it is well-known that, e.g., the viability of knockout mutants[42] invariably depends on the media and growth conditions used to test that viability, a point repeatedly recognized and readily appreciable as well from decades of work in genetics with auxotrophic and temperature-sensitive mutants.[43] Thus, if you knock out a gene for the synthesis of a specific required compound which you then supply in the culture medium, the organism may suffer no detectable injury. It was for this reason that, in the peculiar environment of Spiegelman's lab, the Qβ RNA could tolerate the disruption or absolute loss of the genes required for the synthesis of its replicase enzyme, since that enzyme was available environmentally and therefore didn't need to be synthesized. In short, a gene "essential" in one environmental context may be dispensable in another, and vice versa;[44] a knockout may be rescued by genetic or environmental supplementation.[45]

---

[41] I will return to this question below.

[42] That is, a mutant in which a particular gene has been deliberately disrupted or removed ("knocked out") in order to assess some aspect of its function.

[43] Mutants which have unusual nutritional needs or altered temperature ranges for growth compared to their wild-type ancestor.

[44] It might be objected that an empirical or experimental critique of a proposed essence is a category mistake, since an essence may not be

Koonin (2003) later acknowledged and endeavored to approach this problem of the contextuality of what is identified as essential. He proposed that an essential gene be defined as "a gene for which knockout is lethal under certain conditions." But can an essential gene really be conditional? Wouldn't this be to concede that the "essential," for life, is actually accidental? Why continue nonetheless to valorize the essential then? If, following Koonin (2003) and others who explicitly invoke "conditional" (e.g. Sassetti, Boyd and Rubin 2001; Stahl and Stintzi 2015) and "quasi" essentiality (Hutchison et al. 2016),[46] we accept this strangely qualified notion of the essential, then surely *all* genes will turn out to be essential, at least under some conditions – and all the more so

---

empirically or experimentally accessible, as Descartes's famous example of the wax implies. To concede the point, however, is to admit the futility of the essential gene project from the very beginning – and in the process to abandon an opportunity to investigate what is at stake in such a project.

[45] A genetic supplement would be some other gene which can in some way fulfill the "function" of the knocked-out gene; an environmental supplement would be, e.g., the environmental availability of the gene product or of whatever that gene product enabled. Insofar as genes themselves may be environmentally available – which underlies the phenomena of horizontal gene transfer and of viruses – an absolute distinction between genetic and environmental supplementation cannot be made. But this is really only to admit that an absolute distinction between inside and outside cannot be made. And it cannot. In a certain sense, this is already signaled by the bioenergetic requirement that biological membranes be semipermeable. An absolutely impermeable membrane makes of the living thing a closed system, which accordingly has no means to stave off entropic decline; but neither can a living thing with an absolutely permeable membrane stave off that decline, if only because it lacks the identity against which the decline is measured.

[46] As discussed in more detail in the Parodos.

when among those possible conditions organismal competition is included (on this point, see the critiques of D'Elia, Pereira and Brown 2009 and Fang, Rocha and Danchin 2005). Koonin recognized this problem as well, but his attempt to get around it might be mistaken for allegory. The minimal gene set, the inside, can be meaningfully defined only if the environment, the outside, is likewise defined: "the phrase 'minimal gene-set' in itself makes no sense," he wrote, and "... is only meaningful when associated with a defined set of environmental conditions" (Koonin 2003). Must we not ask, however, under what extraordinary circumstances all environmental conditions, physical *and* biotic, can be defined?

The minimal gene-set and essential gene concept are further eroded when the environment is understood to include a temporal component, with a past and a future. Ironically Koonin himself called attention to this complication in an argument defending the essentiality of RecA, a protein involved in DNA repair and recombination, despite the inconvenient fact that *M. genitalium* RecA-knockouts are viable (Hutchison et al. 1999): Koonin (2000) noted that a cell lacking such basic repair capacity nonetheless "clearly is not facing a bright future."[47] Whole genome sequencing of two strains of the aphid endosymbiont *Buchnera aphidicola* has since revealed that they too lack RecA, even without the

[47] An important point is hidden in this qualification. Why are we driven to associate life with futurity, and what should the horizon of that futurity be? Put otherwise, is any living being facing a bright future, or is life that phenomenon which would persist only through the indefinite deferral of a future all too certain and in nowise bright? This emphasis on futurity points to the structure of deferral and difference (or, simply, différance) discussed both in the main text and in footnote 39.

intervention of laboratory mutagenesis (Tamas et al. 2002). Moreover, if the absence of RecA in both genomes reflects a characteristic of their last common ancestor, rather than independent loss in each lineage, then the two strains of *Buchnera aphidicola* have persisted without RecA for at least 50 million years. Must one tremble for their futures? Such pathos is no more justified than an equally wild inference: that all cells have minimal gene sets, if conceived as situated in a fluctuating, uncertain and biotic environment, one whose absolute spatial and temporal borders are no more certain than life's.

Of course, some minimal gene sets are smaller than others. The final list derived by Mushegian and Koonin consisted of 256 genes. Hutchison et al. (1999) used global transposon mutagenesis to estimate that between 265 and 350 protein-coding genes were essential to *M. genitalium*; 478 were identified as indispensable to *H. influenzae* (Akerley et al. 2002), 192 to *Bacillus subtilis* (Kobayashi et al. 2003), between 227 and 241 to different strains of *Streptococcus pyogenes* (Le Breton et al. 2015), and 620 to *Escherichia coli* (Gerdes et al. 2003). In a recent compilation of literature estimates from 33 different Bacteria, one Archaeon and a few Eukarya (Luo et al. 2014), the number of essential Bacterial genes ranged from 105 [for *Salmonella enterica* serovar *Typhimurium*] to 789 [for *Vibrio cholera*]. Even *Prochlorococcus marinus* SS120, with 1884 protein-coding genes, has been put forward as "a nearly minimal oxyphototrophic genome" (Dufresne et al. 2003) – despite its not being even then the smallest known *Prochlorococcus* genome! Meanwhile, Hutchison et al. (2016) required 473 genes for their synthetic "minimal bacterial genome" – but in arriving at that number were compelled to create a new category of genes, those that were "quasi-essential," i.e. "needed for robust growth." But surely in the context of a biotic environment, one in which organisms compete with (as well as aid) one another, robust growth may be more than a merely desirable characteristic.

The variability in these estimates no doubt to some degree reflects the precise question being asked and differences in computational and/or experimental approaches. More interesting than their variability, however, is the consistency with which these studies have overlooked much smaller (and even related) cellular genomes, frequently without comment. Thus, plastid genomes[48] commonly have gene sets half the minimum proposed by Mushegian and Koonin and a tenth of the "nearly minimal" oxyphototrophic genome of their cyanobacterial relative *Prochlorococcus*. Plastid-derived organelles like apicoplasts typically have even smaller genomes, and the smallest incontestably mitochondrial genome, in *Plasmodium falciparum*, is < 6 kb with only 3 genes! The shared ancestry of mitochondria and plastids with α-proteobacteria and cyanobacteria, respectively, is no longer seriously doubted; each type of organelle is cellular, contains genetic material, can synthesize proteins, is enclosed by a lipid membrane and (in almost all cases) can be included on the ribosomal tree of life; and yet organelles are excluded from analyses seeking to define the minimal gene set! What is the justification for such exclusion?

---

[48] That is, the genomes of the eukaryotic organelles typically involved in photosynthesis. Mitochondria, mentioned later, are the eukaryotic organelles typically involved in respiration.

"There is a qualitative difference," wrote Mushegian and Koonin, "between complete bacterial genomes and any sequence, including viral and organellar genomes, that have been available before. However small, a cellular gene set has to be self-sufficient in the sense that cells generally import metabolites but not functional proteins; therefore, they have to rely on their own gene products to provide housekeeping functions." The qualitative difference separating organelles and viruses from cells is a question of the sufficiency of the self. But which self are we talking about, the self which replicates or the self which is replicated? Isn't the real problem here that these two selves have diverged too dramatically? The virus or the organelle gets replicated only by the virus or the organelle PLUS a highly specific environment which we call their host and which itself replicates. That evolution can lead to such a specific, self-replication-enabling dependence Spiegelman's monster, HERV, *Mycobacterium leprae*, *Nanoarchaeum equitans*, *Encephalitozoon cuniculi* and even slave-making ants have already demonstrated. On what ground, then, do we draw the line establishing our qualitative difference – and on what ground do we hold that line? With what justification will we sort Spiegelman's monster, HERV and organelles to one side of the line, and *Mycobacterium*, *Nanoarchaeum*, *Encephalitozoon* and slave-making ants to the other?

For Mushegian and Koonin, the line of demarcation corresponds to the cell-enclosing, cell-defining membrane – a membrane which, from the start, they recognize as semi-permeable. The transport of myriad entities from one side to the other is, for them, unproblematic – so long as not all entities are transported. In particular, the line is crossed (as it were) as soon as the cell's membrane permits (or fails to prevent) the transport of a functional protein. This they justify by invoking cellular housekeeping:[49] even while they have been cavalier (or at least unsophisticated) in their treatment of a cell's environmental context, of its irreducible ecology (< oikos [οἶκος, house]), they privilege a different domesticity, a properly internal housekeeping in terms of which self-sufficiency would be defined.[50] In so claiming, Mushegian and Koonin were not so much making as repeating a classical distinction. For example, Lwoff, in his 1957 address "The Concept of Virus," also specifically linked the virus to the organelle and concluded the former was not an organism because the latter was not: "Cellular organelles are not organisms because they are dependent, because they are parts of a whole which is the organism itself, the ultimate unit of replication." Dependency (or is it exploitation?) disqualifies the organelle (and thus the virus) from organismal plenitude; it disqualifies it from being considered a whole.

Yet, as soon as we recognize that dependency (and hence wholeness) comes in degrees – a recognition implicit in the divergence I have described between the two selves in self-replication – we will again have to decide where to draw, in other than an accidental fashion, our border-establishing lines.

---

49 Of course they never discuss (and perhaps never considered) the metaphorical dimensions of this term. Such a discussion would in any event be inappropriate to a scientific paper, for which the unenforceable prohibition against rhetoric is frequently satisfied simply by banning rhetoric's explicit acknowledgement.

50 But to which self, we might ask, does housekeeping properly belong, the self which replicates or the self which is replicated? And how to make sense of the affluent (or the master ants) who can afford housekeepers?

For if dependency distinguishes organelles and viruses from organisms, how will that same criterion not exclude slave-making ants and obligate parasites, symbionts, and sexually-reproducing organisms (to name a few) from being organisms? Will we not be left ultimately in the amusingly untenable predicament of prohibiting organisms from being organisms? Surely, to take an example posed by the virologist N. W. Pirie in 1937, "a cow is not classified as 'non-living' because its existence in an environment in which grass is the principal foodstuff is dependent on the cellulase in its intestinal flora!" Ought this not, in Darwin's words, "to convince us of our ignorance on the mutual relations of all organic beings – a conviction as necessary as it is difficult to acquire?" (Chapter III, p. 87) For what organism is not dependent on another organism?[51]

Mushegian and Koonin might object that such criticism is too extreme. Unlike Lwoff, their prohibition is not against dependency *per se*, but rather the import of proteins. Cells that

---

[51] Or, put more dramatically, what organism is not a parasite? Joshua Lederberg (1952, pp. 422 – 3) observed something similar (though he left open the possibility of self-sustaining autotrophs): "...[A]ll heterotrophs are genically insufficient and depend for their subsistence on the bio-synthetically expressed genic functions of other organisms. Thus it is possible to formulate a graded series of symbioses as genic interactions, from the co-habitants of a single chromosome, through heterokaryons and plasmids, to extracellular ecological associations of variable stability and specificity. Other writers have commented on levels of cooperation in societal rather than genetic terms, emphasizing the relationships of genetically similar units such as cells, and some have drawn holistic conclusions. Without furthering such extrapolations, our survey of symbioses should show that the delineation of organic units, be they genes, plasmids, cells, organisms, genomes or colonies is a tool of investigation and communication and not an absolute ideal."

depend on other cells for oxygen, simple sugars, complex vitamins or peptides are (on their view) perfectly reasonable as well as widely attested. Similarly, cells that are dependent on (or exploit) others for glycolysis, the TCA cycle and biosynthesis of lipids, cofactors, nucleotides and amino acids are nonetheless acceptable as organisms. Indeed, Mushegian and Koonin's own analyses used the genomes of apparently obligate parasites. But import of a functional protein, though only a qualitative difference, is the basis of an essential distinction – and this despite the fact that several means are known by which bacteria exchange functional proteins (e.g. Heinemann 1999; Lazdunski et al. 1998; Wilkins and Thomas 2000).[52] Are we therefore to understand that protein import is tolerable so long as it is a bad habit but not an addiction? Is the justification that addiction frequently leads to other vices – or other vices to it? In the case of organelles, they feed their habit by promiscuously exporting genes, including genes whose products they still need, to the nuclei of the cells which host them. Such prostitution is insult added to injury: not only are organelles insufficiently self-sufficient, excessively reliant on import, and inappropriately dependent on gene products which they lack and on their genomically-encoded environment, but additionally they have the gall to require the products of genes which they formerly encoded – and to be

[52] Nor is import of functional proteins limited to the bacterial realm. On this score, one might wonder how Mushegian and Koonin would classify an insulin-dependent diabetic, whose biological housekeeping functions require the import of a functional protein (insulin). Perhaps they would argue that insulin is really a peptide, though the distinction between peptides and proteins (based only on size) is acknowledged to be an arbitrary one.

wildly successful! Yet, if we find such monstrous vices exasperating, we should ask why. Is it not precisely that the distinction must be "legislated" lest evolution not make it?[53] Or put another way: what system of justice will enforce the prohibition that we have legislated, if evolution does not? On what if not qualitative[54] differences are evolutionary distinctions founded and on what but these do they founder?

How far can we take this critique of the vain search for undiscovered and undiscoverable biological essences? Juhas, Eberl and Glass (2011) claimed that "essential genes are, by definition, absolutely required for the survival of an organism and are therefore considered the foundation of life." Later, they admitted that this was an assumption: "We believe that it is safe to assume that some genes are absolutely necessary for the survival of any life form." Later still, the assumption became indubitable: "There is no doubt, however, that some genes are absolutely necessary for the survival of any lifeform." This rhetorical slippage in a single paper already indicates that the ground beneath their feet, "the foundation of life," is far from stable, however much they would like to insist otherwise. So, instead of approaching this assumptively and/or dogmatically (as beyond doubt), what if we ask a simple, empirical question: is it true that at least *some* genes are absolutely necessary for life?

Let us consider parasitic Eukaryotes like *Trachipleistophora hominis*, *Giardia intestinalis* and *Entamoeba histolytica*. Long considered amitochondriate (and by some, primitively

---

[53] For this legislative metaphor, see the end of the Parodos, including its footnotes 40 and 41.

[54] Hence, accidental.

amitochondriate), in fact these have remnant organelles of probable mitochondrial (that is to say, Bacterial) descent, though no genomes have been detected associated with them (León-Avila and Tovar 2004; Tovar et al. 2003; Williams et al. 2002). The vigorous debate in which these purported organelles are ensnared revolves not around their exemplarity as genomeless "organisms." This possibility, if true, has been utterly disregarded: to my knowledge, these organelles have no more been considered "organisms" than their Eukaryotic hosts have been considered "environments." Rather, the debate revolves around the status of primitive Eukaryotes and, in particular, whether or not they had mitochondria. Why such single-mindedness marshaled in defense of a wayward Eukaryotic ancestor and origin? What slips past the sentry as a result? What crouches unseen within the Trojan horse?

In the hindgut of the cockroach *Blaberus* sp. are quartered the anaerobic heterotrichous ciliates *Nyctotherus ovalis*. In turn sheltered within *N. ovalis* inside of *Blaberus* are endosymbiotic methanogens (i.e. Archaea). The methanogens inside of *N. ovalis* inside of *Blaberus* themselves are physically associated with larger, hydrogen-producing organelles, the hydrogenosomes; presumably the methanogens avail themselves of the hydrogen to reduce carbon dioxide. To my knowledge, the organismal status of the cockroaches, ciliates, or Archaea has never been challenged, while that of the hydrogenosome has never been asserted. What is the essential difference justifying this distinction? That the cockroaches, ciliates and Archaea can theoretically live independently but the hydrogenosomes cannot? Do we really know that this is true, either that the first three organisms, under some circumstances, can truly live independently[55] or that the hydrogenosomes cannot? The absolutely necessary qualification "under some circumstances," by admitting the central importance of relations, is what really matters here and what decisively tempers the ambitious adverb "independently," revealing it always to have been an exaggeration. After all, as a simple example, what if we take into account competition? Will cockroaches cured of their microbial denizens be competitive in a world of uncured cockroaches?[56] If the answer is no, are we right to dismiss that fact as merely accidental?[57]

Until recently, sometimes contentious debates focused on the ancestry of the hydrogenosome and specifically on its affiliation with mitochondria. Unlike the proposed mitochondrial descendants of *T. hominis*, *G. intestinalis* and *E. histolytica*, only the endosymbiotic ancestry of the hydrogenosome, not its endosymbiotic origin, has been widely disputed (Dyall et al. 2004). The debate is not whether the hydrogenosome derived from an ancient cell, but rather,

---

[55] What would this even mean? To meet Pace's definition, proof of life would require self-replication; self-replication, I have argued, will necessarily include difference; and hence the entity supposedly living independently will necessarily be surrounded by other entities very (but not completely) like it. Does that count as living independently?

[56] This desire to uncover the organism in an abstract and isolated purity from which all accidents have been removed, and especially those accidents which also lay claim to being organisms – such as parasites – is this not another version of a Cartesian project whose seeming triumph is less the discovery of the *cogito* than its liberation of life, at least human life, from embodiment, finitude and death?

[57] Of course, we might also ask how long they must endure before we are satisfied. Will an instant suffice? Here again we re-encounter that emphasis on futurity which points to différance, as suggested in earlier footnotes.

which ancient cell. The discovery of a small (~ 20–40 kb) genome in the hydrogenosome of *N. ovalis* strongly favors descent from mitochondria (Boxma et al. 2005). It is noteworthy that most efforts have failed to detect genomes in other hydrogenosomes (e.g. Clemens and Johnson 2000; van der Giezen et al. 1997) and that hydrogenosomes are commonly described as "genomeless" (e.g. Embley and Martin 2006). That such a designation ("genomeless") should receive acceptance betrays both an underestimation of the logical difficulty of demonstrating absence and, more importantly, a failure to appreciate the significance of a candidate "organism" having evolved away its genome. It begs the question again why hydrogenosomes, along with other organelles, are ignored in analyses seeking to determine not just the minimal gene set, but the set of genes essential for cellular life.

"Conceivably," wrote Koonin (2003), "the absolute minimal set of genes will correspond to the most favorable conditions possible, in which all essential nutrients are provided and there are no environmental stress factors. Under these conditions, an organism might be expected to shed many genes, sticking to the bare essentials." [58] But what are the bare

---

[58] This claim is interesting for several reasons. First, notice that, to find the cell's genetic "bare essentials," "essential nutrients" must be supplied. If those nutrients aren't provided, then the cell will need extra genes (beyond its "essential" ones) for their synthesis. It would seem, then, that the cell has two distinguishable, though related, essences: essential needs (e.g., the nutrients) and essential capacities (e.g., the genes). Indeed, this slippage is found in the original paper by Mushegian and Koonin: after identifying the genes in common between the two parasites and removing those specific to their parasitic lifestyles, they then *added* genes not found in both organisms in order to supply what they called "essential functions," functions which they

essentials? "Undiscovered and undiscoverable," as Darwin wrote of the essence of the term species? Or, at least in an environment understood as necessarily biotic as well as physical, are the bare essentials, quite simply, nothing? Is this really beyond doubt?

---

posited both organisms required but which probably were carried out by different genes in the two organisms. The assumption, then, is that essential needs and essential capacities are both independent of each other *and* have a one-to-one correspondence. Yet this is open to a critique similar to that of civil society in Rousseau's *Second Discourse*. First, is it clear that we can always distinguish what is desirable for an organism from what it absolutely needs? Might not this distinction in fact depend on context? Second, might not the acquisition of a new capacity incur new needs? That is to say, essential needs and essential capacities are not necessarily independent of one another, and the newfound ability to fulfill one need may create additional needs.

A second aspect of Koonin's claim meriting comment is his emphasis on the absence of environmental stress factors (if such an absence is really conceivable). The experiment that led to Spiegelman's monster suggests a more complicated scenario. There, despite all "essential nutrients" being provided, genome (or at least RNA) reduction was nonetheless *in response to* stress factors – or at least to circumstances which favored faster over slower replication. Do such circumstances count as an *environmental* stress factor? Is the stress factor simply exogenous to and imposed upon the being in question? One might be tempted to say yes, since selective pressure for faster replication was levied by the frequency of transfer from one test tube to another. But this is only part of the story; that frequency of transfer exerted selective pressure only to the extent that variability existed in a replicating population. That variability in turn is another expression of the equivocation implicit to self-replication: clearly the replicated self was not identical to that of which it would be the replica, otherwise whence did the variability derive? Thus from Spiegelman's monster we might conclude, first, that absent stress factors, a genome would be expected to change randomly, not necessarily becoming smaller; and second, that those stress factors cannot *a priori* be designated as wholly exogenous (environmental) or, for that matter, endogenous, but rather are between the two.

In making this argument, the challenge I am trying to pose is not to invite as somehow clarifying an understanding that organelles – even those which appear to have lost the entirety of their genomes – are really organisms. Such an understanding would simply be recuperative, a way of kicking the same old can down the same old road. Rather, I am trying to expose a critique already legible in the field of biology and represented here by organelles – but also by symbionts, parasites, slave-making ants and monsters. This critique challenges an essentialist view of life and its requirement for an absolute, unequivocal and non-accidental distinction between the organism and the (always biotic) environment. But these, I argue, have already been inextricably confused, whether in the equivocation of the term "self" in "self-replication" or in the (related) dialectical entanglement of self-replication and evolution. They are confused as soon as we wonder why an organism's genome should necessarily be contained by it, or as soon as we question our ability to distinguish an inside and an outside absolutely. If, in what will become an instituting recursion, we assume (or inherit) definitions for these terms (whether "inside" and "outside," or "organism" and "environment"), the breach of those definitions and the manner by which they are put into play are marks of evolution, marks already inscribed, as it were, in the space separating the self which is replicated from the self which replicates, or in that space, the *semi-permeable* membrane, by which biology would demarcate self from other.

In the case of mitochondria and hydrogenosomes, an ancestor of the α-proteobacteria adapted to a (biotic) environment by relinquishing to that environment most (or all) of its genome, thereby making those gene products environmentally available and the corresponding genes cellularly dispensable. In this and in its subsequent adaptive "reduction," the ancestor was not unlike Spiegelman's monster or HERV-H elements. In all of their cases, the "self" that got replicated diverged dramatically from the "self" that did the replicating; and if that divergence was an accident, it was a necessary and unavoidable one, the abyssal crack that has always already existed in the whole that would be life. Yet, rather than understanding this as a positive fact about life of which our understanding must take account, we instead have been at pains to re-impose the categories that these biological misfits flaunt, even if we must do so arbitrarily. By that logic, however, the question becomes whether, at some point, an organellar ancestor relinquished or, as with an undesired citizenship, renounced life. It becomes the admittedly perplexing question, can life evolve away its own lifeness?

But such an approach is misguided, especially to the degree that it invites only an affirmation or a negation as an answer. Such an answer will only be issued by decree; it will be legislated as by a body politic itself understood by analogy to the living.[59] Just as with the sterile debate about whether, for example, viruses are or are not life, overshadowed will be the more instructive and important question: why can't we decide?

---

[59] The obvious example is Hobbes' Leviathan. My language here and earlier has often taken its direction from moral and political lexica – and not coincidentally. Perhaps inspired by Nietzsche (see aphorism 262 of *Beyond Good and Evil*), I am trying to suggest that essentialist discourses about life end up secretly being moral, political or, if I were to speak most broadly, aesthetical. As with a judgment of taste, they would command the universal assent of mankind – and not get it.

Or, rather, why can we only decide? If life, by unavoidably differing from itself (i.e., by evolving) defers itself (i.e., self-replicates), does it not reveal as a Procrustean bed every category to which it would be fitted?

IV. Conclusion: In what way is life?

In writing this chapter my goal has not principally been to point out the barrenness of the essential gene or minimal gene set concepts or to dismiss Norm Pace's attempt at a definition of life. In fact, both of these projects strike me as strangely productive, not least for a mistake that they share: the idea that life is effectively modular and separable from a context that is itself ultimately reduced to merely physical and chemical conditions. Such an understanding of life I call "essentialist," not only because of its frequent recourse to essentialist language (as obviously with the "essential gene" concept) but also because of its dismissal of accidents as ultimately irrelevant to life. On the contrary, those accidents, which include the myriad relations of living beings to one another, are on my view constitutive of life, or at least of the striving for life that life is. Life is everywhere bounded by such relations; life is inextricably contextual. To try to understand life outside of context is to forget that there's no outside for life to be outside in; the "neutral" space of the laboratory is no less a context than the great outdoors. When an understanding of life is reached which ordains instead that some specific characteristic of life be recognized as essential, and which in so doing isolates life from its accidental context by privileging a particular, though unconfessed, context, that understanding will foreseeably be subject to eventual breach, for example in a context which favors the loss of that characteristic. Indeed, if one ferrets around especially at the disciplinary margins, one may find those breaches have already occurred, though they will, as a rule, not be recognized in these terms.

As an example, even as an allegory, I have tried to describe one such breach involving the loss of genes and genomes, not only by viruses but by cells – cells moreover which we accept as descendants of ancient, putative organisms. Such examples seemed especially choice-worthy in a reductionist age all-too-prone to exalt DNA and genomes as life's ultimate key and long-guarded secret (as discussed in the Parodos). But similar dynamics should be expected for other aspects of living systems besides genomes. For example, by way of analogy, we might be tempted to ask whether viruses emerged in a context favoring loss of metabolism. In a (biotic) environment where metabolism, in one form or another, is no less available than replicases were to Qβ's RNA, why maintain a redundant ability to metabolize? The situation is comparable to the phenomenon of herd immunity, which permits some fraction of a population to avoid being immunized while that fraction still reaps the benefits of immunization.

As attractive as such an argument might seem, however, it runs a risk of falling back into the very assumptions ostensibly being critiqued. In particular, it ends up treating categories like "cell" and "virus" as fixed, as defined, as reducible in some undisclosed way to an immutable essence. It fails to ask whether "cells" and "viruses" – both necessarily in quotes because we don't actually know what the words mean – have always been distinguished in the way that they are now. It fails to consider that their distinctness now is both a matter of degree,[60] on the one hand, and on the other a reflection of their extensive opportunities to coevolve. To what extent did "viruses" and "cells" jointly shape each other and to what extent are they still doing so?

In writing this, it is necessary to call particular attention to the unavoidable mediation of language, not only with regard to the words "virus" and "cell" but with regard to the word "life," which is no less under erasure than the others. These words serve as tags for something that escapes them; it is by them that we would grip Proteus. What one has to understand is not that we succeed in gripping Proteus, but that we fail and can only fail: these words serve as tags for something that *escapes* them. To instead understand my argument, not in linguistic terms, but in historical ones – as if history can achieve what life cannot and be outside of context – is to not understand my argument at all. To seek evidence of ancient "viruses" and ancient "cells" and to search for their origin(s), as if those categories are impervious to the play of evolution and self-replication, of difference and deferral, of *différance*, which has shaped them into the forms we recognize now, is to plummet right back into an essentialist perspective of life. Essentialism is not evaded but assumed by historicizing, by the definitional gesture which discovers beginnings and ends by imposing them.

---

60 This is already suggested by the fact that Lwoff, Mushegian and Koonin, in passages already considered, specifically related organelles, not to the "proper" cells from which they are considered to have descended, but to viruses.

Such mistakes are already in evidence in the sometimes vitriolic debates, not only about the origins of viruses[61] but especially about how to categorize them (e.g. Moreira and López-García 2009 including comments and responses; Villarreal and Witzany 2010). Are viruses life? Do they belong on a "tree of life"? Such categorization can only be adjudicated, whether by judge or by jury; lost in the stir is that the adjudication, whatever its decision, in fact offers no deep insight into the unfolding biological and evolutionary drama in which viruses have more than one part. It does, however, risk obscuring that drama by essentializing what are and can only be provisional categories. From within a dynamic (call it "evolutionary," though "textual" would be better) which critiques any particular characteristic aspiring to essentiality in the starkest way (by elimination), we can expend endless energy debating whether the resultant being meets the criteria of definitions which pre-existed it; but in the process we must not lose sight of the only positive phenomenon before us – our *failure* to grasp precisely that for which we reach – and what that failure is telling us about a category precariously always in play. A category, then, which is not a category, though we call it "life."

---

[61] See, for example, Rice at al. (2004), whose demonstration of the antiquity of viruses requires them to postulate a particular protein fold as the "innate 'self'" (p. 7719) common to the viruses in question and a presumed ancient progenitor. Three facts are notable: first, this innate 'self' is not common to *all* viruses; second, it is shared only in degree, a question adjudicated according to some unspecified index of morphologic similarity; and third, the 'self' which they postulate begins in scare-quotes, presumably to indicate its rhetorical status, but it is soon released from those confines.

And what is it telling us? It is tempting to propose that the essentialist understanding of life be replaced by a relational one, with those relations always already including other beings striving for life. We would then say that the struggle for life is always already a struggle with life, and we would call attention to the ambiguity of the prepositions "for" and "with." Were we to go that route, we would encounter the foreseeable objections of those who cannot help but postulate the origin of life, not seeing in that postulate, much less the definite article, precisely the essence-based thinking which I am endeavoring to critique. To such an objection, we would note that life understood as relational, as a sort of marionette composed solely of a seemingly infinite number of strings by which it is connected to, controlled by and controls only other marionettes similarly constituted, has no discrete origin. On this view, every biotic saltation occurs in a biotic context, a context of other beings striving for life, against the backdrop of which striving the leap became possible.[62] Rather than

---

62 This is again not in itself an unheard-of claim. According to Morowitz (1968), "Traditional biology has tended to concentrate attention on individual organisms rather than the biological continuum. The origin of life is thus looked for as a unique event in which an organism arises from the surrounding milieu. A more ecologically balanced point of view would examine the ecological cycles and subsequent chemical systems which must have developed and flourished for a considerable period before anything resembling organisms appeared" (pp. 55 – 6). A similar appeal to ecology and biological communities can be found in the emphasis on a community of "pre-cells" or of "progenotes" central to the origin-of-life theories of Baross and Hoffman (1985) and Woese (1998), respectively. Obviously I am sympathetic to all of these claims, though I don't consider them to go far enough – in particular, because they do not criticize the assumption of an origin.

adjudicating whether the resultant entity deserves to be dignified as "life" – fitting life to the bed, as it were – we should consider how every such leap not only reflects its context but alters it. Its neighbors may respond by leaping similarly (adoption; they then become recognized heirs); but they may also respond by adapting, in a mode which from our perspective looks "exploitative" or "dependent" and which may accordingly fall short of "life's" new-found dignity. Very well; the recognition of only the first response as legitimate at least has the virtue of confessing that what "life" is is answered only be decree.

Implicit to that confession, however, is the acknowledgement that the word "life" has been, and will continue to be, in play, precisely as a result of its relational "essence," of its composition by way of accident, of its inextricable contextuality, of its paradoxical nature to strive to be what it has yet quite to be – itself. And so we come, with this seemingly postmodern formulation, to the very threshold of intelligibility. Yet what appears as intelligibility is really a predicament and, in fact, an absolute failure. For what is it that we are on the verge of grasping? In our hands we would clutch relations, accidents and contexts over and above essences; and so long as we never wonder how it is that we think relations, accidents and contexts – so long as we never suspect that these terms may in some secret way invoke essences – all will be well. Let life, then, be a relational process – the metaphor of flow is commonly used, as by Woese (2004). Smuggled in with that recognition – a recognition by way of metaphor, like an ancient sea-god grasped as a flame – is a second one: that life *is*. Our efforts to pick apart the terms bookending the question "what is life?", our critique of its whatness and our attempt to take back the givenness of its lifeness, fall short to the degree that the "is" remains bracketed and deferred. Everything which we would critique and rethink is already implied in that "is."

Yet, surely it is too much to ask that life be thought as that which is not *quite*.[63] What can it possibly mean for something to strive to be what it has yet *quite* to be, itself? This is a problem not just of being but of meaning; not just of the meaning of being, but of the being of meaning. It is a problem of language, or especially of writing, one which certainly cannot be confined to discourses about biology or life. It is a problem of a sign which never rests in the fulfillment of its signification, not even as its own "self." It was this which was legible already in the equivocal term "self-replication" – a divided self which, by dividing, replicates itself as something other than itself. Hence the failure of "life" to be either whole or complete – not because it is defined relationally, but because it isn't *quite*. Because not only the question of being, but being itself, has been deferred. That failure to be (fully) (present) is as threatening to the category of relations as it is to that of essences. The terms one will have used, at first without caveats, then more cautiously, then explicitly provisionally – will be revealed, all along, to have been under erasure. The "biotic environment" which I have postulated as having always already been there turns out to have always already been there

---

63 One thinks of Pascal's fragment Vanité (No. 33/38): "Le présent n'est jamais notre fin: le passé et le présent sont nos moyens; le seul avenir est notre fin. Ainsi nous ne vivons jamais, mais nous espérons de vivre..." (The present is never our end: the past and the present are our means. The future alone is our end. Thus do we never live but hope to live...)

almost. Even now. "Life," "environment," "inside," "outside," "relation" and "essence" – we wrote a lot without knowing what any of them were or meant. Why? Because they didn't *quite* mean. Because they were not – not *quite*.

We have always been dealing with fragments. But whose? Life's? Or being's?

Chapter Two, on Evolutionary Biology
Being Fragmented: On the Imperfect Book of Life

"What writing itself, in its nonphonetic moment, betrays [trahit], is life. It menaces at once the breath, the spirit, and history as the spirit's relationship with itself. It is their end, their finitude, their paralysis. Cutting breath short, sterilizing or immobilizing spiritual creation in the repetition of the letter, in the commentary or the exegesis, *confined in a narrow space, reserved for a minority, it is the principle of death and of difference in the becoming of being."*[1]

Derrida (1997), p. 25 (pp. 40–1 in the original French).

I. Introduction

Supposedly in recent years evolution has at last fallen within the purview of human power and governance ("We now have the power to start controlling evolution," synthetic biologist Craig Venter said in a 2015 interview; Gardels 2015). Evolutionary biology is being re-understood as an applied science (e.g., Bull and Wichman 2001, Carroll et al. 2014) charged minimally with the "control of unwanted evolution" (Carroll et al. 2014) and, concordantly, with evolution's management ("we can either choose to manage evolutionary process or not," claimed UCLA evolutionary biologist Thomas Smith in Hewitt 2014); synthetic biology has emerged as a field in which, as Venter told an National Public Radio audience in

[1] One must be careful to hear the word translated as "betrays" in the sense both of deceptive disloyalty and of divulgence, even revelation, as of an innermost secret.

2010, we can finally "create new species to do what we want them to do, not what they evolved to do"; and conservation biologists would, in some cases eagerly, accept the mantle of administering not only extinction, but de-extinction and evolutionary potential as such (e.g., Sherkow and Greely 2013; Zimmer 2013; Jørgensen 2013; Seddon et al. 2014; see also the Parodos and Chapter 3). Yet, if this seems like the threshold of a brave new world, a closer inspection will reveal more than a few wrinkles. After all, this was territory clearly staked out by the eugenics movements of the early 20th century. It is, furthermore, the implicit and deeply problematic assumption underlying Darwin's division of selection into methodical and natural branches. And, as Darwin himself points out, the seeming manipulation of evolution by human beings has marched in near lock-step with the very unfolding of civilization itself: for, if we think genetically-modified organisms are the fraught innovation of modern times, it is only out of a curious forgetfulness of the "high antiquity" (in Darwin's words) of the "methodical practice" of the principle of selection, for which Darwin gives as evidence an ancient Chinese encyclopedia,[2] the book of Genesis, Pliny and the

[2] Surely it is too much to hope that this is the same encyclopedia to which Borges referred in "The Analytical Language of John Wilkins" and returned to again by Foucault in the early pages of The Order of Things. Borges' critique of Wilkins and of his conception of language (and through it of the universe) is not unrelated to the concerns of this chapter; the "applied evolution" projects I criticize rest upon the albeit veiled presumption of a universal language much like that which Wilkins postulated.

practices of 'savages' from South Africa to the poles (On the Origin of Species [henceforth The Origin], Ch. 1, pp. 33-4).[3]

More interesting than a merely renewed rumination on the purported human control of evolution is to examine this forgetfulness. If human beings have wielded evolution towards their own ends for so long, how is it, and why is it, that we keep forgetting that we have done so? What kind of wielding would apparently so forgettable a practice be? In Darwin the pursuit of this question very rapidly leads to problems of agency – not only whether nature, but whether human beings, can truly be selective agents. As discussed in more detail in Chapter 3, Darwin's deployment of unconscious selection as a middle term connecting the unintentional agency of nature with the intentional agency of man allows him to successfully rationalize the trope that nature selects; but it does so at the expense of revealing human selection as *also* secretly figurative, a gesture portending both more and less that what it is. Human agency is never so complete that methodical selection is untainted by its unconscious and, according to Darwin, more important counterpart, as Darwin's near-religious pairing of the two terms implies; and so human agency inevitably falls short of its aims precisely to the extent that human selection, with equal inevitability, exceeds them.

This paradoxical – or rather, *ironic* – combination of success and failure, agency and passivity, is echoed repeatedly throughout The Origin. It is – *irony* is – a rhetorical motif of that text. In it, taxonomic categories like species are at once well-defined and incapable of definition. Origins are found

---

[3] All quotations, unless otherwise noted, are to the first edition. I have silently substituted American for British orthography.

only to recede into a more distant past. Likeness is always the metonym of common descent – except when it isn't. The possibility of absolute perfection is at once insisted upon and annihilated. In short, our forgetfulness is more than an incidental shortcoming; it is the reflection of a curious and seemingly self-consuming logic which, in always interrupting itself, mirrors the constantly interrupted natural phenomena which it describes. In a few words, that interruption, that permanent parabasis,[4] reveals the natural world depicted by Darwin's text – but also the text itself – as a phenomenalization, or rather allegorization, of irony.

Irony here is not incidental or avoidable but necessary. As I will argue in this chapter, underlying that necessity is a metaphysics of presence which, on the one hand, Darwin constantly assumes, and which, on the other, his text no less constantly calls into question. Thus he insists upon a seemingly benevolent and sometimes capitalized Nature tirelessly working toward the improvement of her charges; but that improvement, he also describes, is the accomplishment of Nature in a ruthlessly, and indeed lethally, exacting mode from which, in the end, not a single one of her charges will have been exempted. "Man selects only for his own good," Darwin writes in Chapter IV, "Nature only for that of the being which she tends." But Darwin never asks himself the question of the meaning of "being" here, nor does he ever explicitly confront how the good of that being is achieved by both preservation and extinction. He cannot help but assume, like the western tradition of which he is heir, that the question

---

[4] I am alluding to Schlegel's famous definition of irony.

of the meaning of being is to be answered as "presence";[5] but what he describes always ends up substituting something else precisely in the place where unadulterated presence would stand. Life as he depicts it is never quite and never yet itself;[6]

---

[5] Cf. Heidegger, Being and Time (1962, p. 47): "In other words, in our process of destruction we find ourselves faced with the task of Interpreting the basis of the ancient ontology in the light of the problematic of Temporality. When this is done, it will be manifest that the ancient way of interpreting the Being of entities is oriented towards the 'world' or 'Nature' in the widest sense, and that it is indeed in terms of 'time' that its understanding of Being is obtained. The outward evidence for this (though of course it is *merely* outward evidence) is the treatment of the meaning of Being as παρουσία or ουσία, which signifies, in ontologico-Temporal terms, 'presence' ["Anwesenheit"]. Entities are grasped in their Being as 'presence'; this means that they are understood with regard to a definite mode of time – the 'Present.'" Likewise Derrida in Of Grammatology (1997, p. 12), implicitly and then explicitly with reference to Heidegger: "We already have a foreboding that phonocentrism merges with the historical determination of the meaning of being in general as *presence*, with all the subdeterminations which depend on this general form and which organize within it their system and their historical sequence (presence of the thing to the sight as *eidos*, presence as substance/essence/existence [*ousia*], temporal presence as point [*stigmè*] of the now or of the moment [*nun*], the self-presence of the cogito, consciousness, subjectivity, the co-presence of the other and of the self, intersubjectivity as the intentional phenomenon of the ego, and so forth). Logocentrism would thus support the determination of the being of the entity as presence. To the extent that such a logocentrism is not totally absent from Heidegger's thought, perhaps it still holds that thought within the epoch of onto-theology, within the philosophy of presence, that is to say, within philosophy *itself*. This would perhaps mean that one does not leave the epoch whose closure one can outline."

[6] William Paley's Natural Theology (1833) famously compares living beings to a watch found in a heath. But in a strange way, life as described by Darwin is also a timepiece. In Chapter V, "Laws of Variation", there is a curious slippage between these "laws" – the effects of external conditions, use and

disuse, acclimatization or habit, correlation of growth and reversion – as providing the variations from which nature selects, and as *alternatives* to Natural Selection. Thus, for example, p. 133 ("we cannot tell how much of [variation] to attribute to the accumulative action of natural selection, and how much to the conditions of life"), p. 135 ("In some cases we might easily put down to disuse modifications of structure which are wholly, or mainly, due to natural selection"; see also p. 147: "I see hardly any way of distinguishing between the effects, on the one hand, of a part being largely developed through natural selection and another and adjoining part being reduced by this same process or by disuse, and, on the other hand, the actual withdrawal of nutriment from one part owing to the excess of growth in another and adjoining part"), p. 141 ("How much of the acclimatization of species to any peculiar climate is due to mere habit, and how much to the natural selection of varieties having different innate constitutions, and how much to both means combined, is a very obscure question"), p. 144 ("I know of no case better adapted to show the importance of the laws of correlation in modifying important structures, independently of utility, and therefore, of natural selection, than that of the difference between the outer and inner flowers in some Compositous and Umbelliferous plants"; see also p. 146: "I do not doubt that some apparent correlations, occurring throughout whole orders, are entirely due to the manner alone in which natural selection can act") and p. 153 ("the struggle between natural selection on the one hand, and the tendency to reversion and variability on the other hand, will in the course of time cease"). As the last example begins to suggest, this slippage will be rationalized ultimately by understanding these laws and Natural Selection as operating on different temporal scales. Darwin's claim is that Natural Selection will eventually "more or less completely, according to the lapse of time, [overmaster]" (p. 158) and co-opt all these others. But the lapse of time at issue will require Darwin to look, as he admits in the next chapter, "not to any one time, but to all time" (p. 179). The result is a living being who is both chronological misfit and historical palimpsest, with "the embryo... left as a sort of picture preserved by nature, of the ancient and less modified condition of each animal" (p. 338) or "organs now of trifling importance" though formerly "of high importance to an early progenitor" (p. 195) – for example, upland geese and frigate birds with uselessly webbed feet, and myriad equids bearing, literally, the stripes of their ancestry. "Hence

and as he tries to grasp this slick, squirming Proteus on the constant assumption that it is, after all, present, what ends up in his hands is always a little different – offspring resembling, but not exactly, their parents; a monstrosity eliding into a continuum of variation; a variety that, seen in a different light, is a species; and a tree of life which is also the Tree of Life and, at the same time, anything but a tree.

The basic question of this chapter is "What, according to The Origin, is the being which Nature tends?" Although Darwin never directly confronts this question, the strangeness of that being everywhere opposes him and governs his text. It forces him to describe life in terms of an explicitly referential structure; to substitute synchronic observations for diachronic inferences, and vice versa; to emphasize life's contextuality over and against its essence; to relinquish both agency and origins; and ultimately to undercut the concepts, and the very concept of concept, on which his edifice would rest even as instead it trembles. Darwin's far-sightedness, it turns out, is indistin-

---

every detail of structure in every living creature (making some little allowance for the direct action of physical conditions) may be viewed, either as having been of special use to some ancestral form, or as being now of special use to the descendants of this form – either directly, or indirectly through the complex laws of growth" (p. 200; just before, he specifically relates these observations to "natural selection, subjected (!) formerly, as now, to the several laws of inheritance, reversion, correlation of growth, &c."). What has Darwin described, if not a living being subject to an hour hand, a minute hand and a second hand – each one shaping it differently, though ostensibly toward a single end? Yet, has he not also described a being whose time is out of joint, a being burdened with a seemingly useless past, anxiously projecting itself into an uncertain future, and meanwhile unable to catch up with the present which it supposedly inhabits?

guishable from his blindness.[7] Thus the man who, echoing Polonius, "can see no difficulty in a race of bears being rendered, by natural selection, more and more aquatic in their structure and habits, with larger and larger mouths, till a creature was produced as monstrous as a whale" (Chapter VI, p. 184), who "can see no difficulty in natural selection making an occasional habit permanent" (Chapter VII, p. 219), who, with regard to insect castes "annually born capable of work, but incapable of procreation,... can see no very great difficulty in this being effected by natural selection" (Chapter VII, p. 236) and who is even capable of seeing without difficulty what is not there ("For myself, I venture confidently to look back thousands on thousands of generations, and I see an animal striped like a zebra..." Chapter V, p. 167) is *also* the man keenly attuned to the legitimate challenges a theory faces from what cannot be seen: on the theory of independent creation, he observes over and over again, he "can see no explanation" for a variety of phenomena (e.g., the hierarchical structure of classification [Chapter IV, p. 129] or why remarkably developed organs are also highly variable [Chapter V, p. 152]), while his own theory is also acknowledgedly troubled by what cannot be seen, the missing intermediate forms, the want of absolute perfection,[8] and the blanks of the geological record.

[7] I shall return to the significance of sight especially in Chapter 4.

[8] Or at other times the want of that want, for Darwin can't decide whether perfection or its absence is the more troubling – or marvelous. Thus: "Nevertheless, such a conclusion [that species had descended, like varieties, from other species], even if well founded, would be unsatisfactory, until it could be shown how the innumerable species inhabiting the world have been modified, so as to acquire that perfection of structure and coadaptation which most justly excites our admiration" (Introduction, p. 3); yet later we

And so his theory unfolds much like what it would describe: that which can be seen – the (for now) preserved – is articulated against and by that which cannot be seen – the absent, the blank past and the unknown future, the extinct and the generations to come – as if the one were letters and the other spaces. As if what he is describing in writing is itself a form of writing. And, by no means coincidentally, it is to writing as a metaphor that he can't help but repeatedly turn, even if he doesn't understand why.

But that is what we shall endeavor to understand in this chapter, as the inescapable correlate of understanding the being which Nature tends.

II. Imperfect origins: The place of apology

Darwin has barely begun On the Origin of Species when he apologizes. "This Abstract" – for, we must never forget, that is all it is, an abstract of an argument yet to come, an argument moreover that in a single form never comes, although Darwin assures us that the "work is now nearly finished" and will require only "two or three more years to complete" (so long as he doesn't die, his "health is far from strong"; Introduction, p. 1), yet in lieu of its promised completion, that argument to come gives forth instead to a wide variety of related works, each taking its specialized place in the natural economy of the overarching theory and thus marking the enormous range that theory would inhabit were it finally to be perfected into its

---

are consoled that we "need not marvel at some instincts being apparently not perfect" (Chapter XIV, p. 475). "The wonder indeed is, on the theory of natural selection, that more cases of the want of absolute perfection have not been observed" (Chapter XIV, p. 472).

promised presence – "This Abstract, which I now publish, must necessarily be imperfect" (p. 2). Indeed he is absolutely right; demonstrating the necessity of that imperfection, that incompleteness, will be a project of this chapter. Before that, however, it is worth noting what it is that stands out to Darwin as especially indicative of the present imperfection of his abstract for a text eventually to come. It is its want of references and authorities. "This Abstract, which I now publish, must necessarily be imperfect. I cannot here give references or authorities for my several statements..." A striking inability, on many counts, not least his acknowledgement that his statements do not have a *single* or *ultimate* authority, author or origin, an absence that foreshadows a similar lack in the Book of Life[9] from beginning to end: for, he confesses in Chapter I (p. 40), "a breed, like a dialect of a language, can hardly be said to have had a definite origin" – and if not breeds, then not varieties; and if not varieties, then not species; and suddenly the definite article in his title becomes at once ponderous and ponderable – while, 450 pages later, on the abstract's last page, the first edition famously concludes with a Creator originally absent, a blank overdeterminedly filled in every subsequent edition: "There is a grandeur in this view of life, with its several powers, having been originally breathed [*by the Creator*, every subsequent edition appends] into a few forms or into one" (Chapter IV, p. 490).

In lieu, in the place, of these missing and notably plural authorities, what shall we rely on? "This Abstract, which I now

---

[9] Or, in the words of Bacon quoted as an opening epigraph to The Origin, "the book of God's works" (p. ii).

publish, must necessarily be imperfect. I cannot here give references or authorities for my several statements; and I must trust to the reader reposing some confidence in my accuracy." Darwin trusts us readers to trust him. He rests his trust in us – who for him, as he writes, are yet to come – that our trust will rest in him – who for us, as we read, has already come. It is a curiously deferring logic, one that leaves a place blank in the promise that it either will be filled or has been filled. But what would fill this blank? The answer is strange, for it isn't simply Darwin or a reader; it is Darwin *taking the place of* the references and authorities he can't for purely incidental reasons, only temporarily and at that particular moment provide – even though the imperfection is also, as he acknowledges, *necessary*; and it is a reader, also a placeholder, who, Darwin anticipates, accepts in advance the legitimacy of his place-holding, out of trust that the references and authorities will come, or at least in principle could come, and who therefore accepts Darwin, of course only provisionally, as a reference to them and for them. In short, the lack of authorities, and ultimately the lack of a single authority, drives us to reference, albeit a reference which we must also connect to deference: a reference not to what is there or even somewhere, but to what is absent, what could come or may once have been, but to which, in and because of its present absence, and in its capacity as *an* but not *the* authority, we can only defer.

In this brief apology is the referential structure of the entire argument and that the entire argument is. For The Origin is in an important sense a treatise on biological rhetoric, on how to read and, if always by laws but dimly seen, on how to regulate biological tropes. Always it is a question of relating what is there, which in itself is never sufficient, to what is not. When we observe that this variety is like that variety, and thus when we strictly limit our metaphor to objects contemporary with one another, it is nonetheless not what it seems, "the mere putting together and separating objects more or less alike" (Chapter XIII, p. 420); rather it is our fumbling approach to that "hidden bond which naturalists have been unconsciously seeking" (*Ibid.*), community of descent – a bond whose revelation is the accomplishment of a posited rule that likeness reflects shared (but differently shared) similarities to a common and no longer extant parent. Thus, as powerful as synchronic comparisons are, they become meaningful for Darwin only when interpreted diachronically, that is to say, with respect to what is no longer. Or with respect to what is not yet. For the same gesture can be projected into the future no less than cast into the past;[10] it can even be cast

[10] Darwin's text, as we might expect, both claims this and enacts and thus represents it. Hence not only his attempted and, he acknowledges, inadequate gesture to recognize the contributions of the past – those authorities whom he would reference if he could but, lacking that, whom he must be the reference for – but also the claim he makes on the future, both in the trust that he exacts of it and in his hopes for the support they will give his intellectual offspring: "A few naturalists, endowed with much flexibility of mind, and who have already begun to doubt on the immutability of species, may be influenced by this volume [i.e. The Origin]; but I look with confidence to the future, to young and rising naturalists, who will be able to view both sides of the question with impartiality" (Chapter XIV, p. 482). Consider also the future geologist (p. 299) whom he invokes in Chapter IX. Finally, while Darwin typically refers to "inheritance with modification" (e.g., Chapter IX p. 350), he sometimes also refers to "generation with modification" (e.g., in his discussion of Wallace's "ingenious paper" in

into a future past.[11] Perhaps that future past would be the most rigorous temporality on which Darwin could insist, all the more for the grim fate he openly anticipates for all extant beings and species; yet he instead seems inclined to privilege a simple future, even if he must extrapolate it back in time to generate a theory, not of altered ancestors, but of modified descendants. He cannot help but describe the motor of that theory in explicitly rhetorical terms:

"I should premise that I use the term Struggle for Existence in a large and metaphorical sense, including dependence of one being on another, and including (which is more important) not only the life of the individual, but success in leaving progeny." (Chapter III, p. 62)

Within a metaphysics of presence, it might be surprising that the life of the individual is *inadequate* for understanding the struggle for existence, even if seemingly it is precisely only that individual life whose existence is in question. But that's the rub: its existence isn't really in question. In fact, its existence is insufficient to itself; and this individual's ultimate fate can be known absolutely. Its impending absence already infiltrates its fragile and incomplete presence; and that infiltration, which is the same as that fragility and incomplete-

---

Chapter IX, p. 355), though these terms have subtly different temporalities: the first is a bequeathal of the past to the present, the second a present projection into the future.

[11] For example, Darwin anticipates a "future historian" who "will recognize [Lyell's Principles of Geology] as having produced a revolution in natural science..."

ness, drives us beyond the individual to its mutual relations with other individuals, on the one hand (and this will be the subject of prolonged discussion in the second part of this chapter, as it was already in Chapter 1), and on the other, to its future likenesses, its descendants. These metaphorical auxiliaries arrive, if they arrive, in the nick of time to supplement a presence revealed as all too insufficient – even if it is their arrival which reveals that. It is, after all, "the very process of perfection through natural selection" by which earlier forms "will have been supplanted," Darwin writes in Chapter VI (p. 182); while, two chapters prior (Chapter IV, p. 119), he acknowledges that "[t]he modified offspring from the later and more highly improved branches in the lines of descent, will, it is probable, often take the place of, and so destroy, the earlier and less improved branches."

Of course these Oedipus-like descendants are also not sufficient unto the day. Having taken the place of, and so destroyed, their parents – for want of, among other things, space – they await the same fate; no more can they exhaust their own representational potential in being merely themselves than could their forebears. Thus they too refer beyond themselves, both to the parents they eliminated and to the promise and threat of their own offspring. But their references beyond themselves are not limited to the past or to the future. According to Darwin, they additionally refer, as also their ancestors did and their progeny will, to something else: and that something else is their *place*, a term Darwin invokes with such regularity that one almost forgets that he never explains it. It is a question, for Darwin, not only of offspring taking their parents' place (e.g. p. 280), but more generally of "occupying," (e.g., pp. 82, 102, 113, 116, 119, 412) "filling (up)" (e.g., pp. 81, 82, 102, 121, 177-8, 315), "fitting" (e.g. pp. 157-8, 183, 199) or "seizing" (e.g., pp. 81-2, 102, 112, 113, 119, 186, 319, 428) a place in Nature, or in the economy of Nature (e.g. pp. 76, 102, 104, 157-8, 315) or even in the polity of Nature (e.g. pp. 102, 106, 109, 112, 121, 126, 177-8).[12] These beings, the very ones which Nature tends and which replace and take the place of their parents, are placeholders, just like Darwin was with respect to the missing authorities to whom he would refer if only it were possible – if only he were healthy and could count on having the time (p. 1), if only he were not limited by a lack of space ("want of space prevents me..." p. 211; see also, e.g., pp. 230, 233 and 248). Temporal and spatial finitude infiltrate all of these beings so radically that their recuperation cannot simply unfurl in time but must additionally lay claim to an enduring place not only occupied, filled, fitted or seized by these beings, but represented by them. And if represented by them, that means – not identical with them. Hence places can be seized as well as better occupied, filled, or fitted; hence one being can take the place of, can represent, and in the same gesture can destroy, another.

This invocation of place both articulates the logic and even language of representation (with one being *taking the place of* another) while also endeavoring to ground it, albeit in the mysterious and, one suspects, figurative realm of place. Place enables the promiscuous representational relay but, in

---

12 They even have "places in a natural classification" which are in danger of being "more or less completely lost" (Chapter XIII, p. 421). One wonders if places (as opposed to placeholders) can likewise be lost in Nature, for Darwin.

accordance with a metaphysical demand for presence, would also end it. Since the beings that nature tends fail to be fully present, they must represent – or so that metaphysics assures us – what *can* be fully present, namely, place. But to accomplish this, the plenitude of place is restricted in what it can be. In particular, place must be purely formal, the biological equivalent to Kant's moral law. As with that law, its content is not and absolutely cannot be integral to it. If it were, one content couldn't be exchanged for another without changing what place *was*; and such a place, no longer eternal and unchanging, could be no place for placeholding. Content furthermore is particular; particularity entails limitation, finitude and hence interpenetration by absence; and if place were so characterized, it would be undistinguished from its placeholder and would therefore be itself a placeholder. If that were the case, however, it would serve no purpose in Darwin's text – and precisely at this moment Darwin's text is insisting on the possibility of purpose, of telos (τελος), of an end to placeholding and thus of an end to representing. In short, a signified that does not itself signify – the ground, the place, on which signification would rest, that is to say, stop.

Yet such a place places Darwin in a bind. It is a problem of counting. "That the number of specific forms has not indefinitely increased, geology shows us plainly; and indeed we can see reason why they should not have thus increased, for the number of places in the polity of nature is not indefinitely great, – not that we have any means of knowing that any one region has as yet got its maximum of species" (Chapter IV, pp. 109-110). How does he know that "the number of places in the polity of nature is not indefinitely great"? How many places are there in what is, after all, a metaphor? How are such places detected, especially since Darwin concedes places may be unoccupied, hence may have no placeholder? Does he see no difficulty in seeing a place with no placeholder? If this is a problem, however, it is superseded by a greater threat, the one to which Darwin is more immediately responsive – the threat of places proliferating without end. For an indefinitely increasing number of places repudiates an ordering τελος; the unconstrained spatial proliferation of signifieds is no less threatening than their perpetual temporal retreat as an origin always in abeyance. A world defined (in the literal sense) by a single origin, that is to say, by an originary, unadulterated and pure presence, is also a world which can tolerate no aborted tower of Babel, no dispersal and scattering of tongues.

But Darwin must also concede that the living world, like his book, is as yet imperfect. This is the other side of the bind and again, counting tells him so, even though it is no clearer now than a few sentences ago how he could count places rather than placeholders. If before he saw no difficulty in seeing empty places – as it were, synchronic absences – now he sees no difficulty in seeing places not yet there – diachronic absences. Hence he is convinced not only that the number of places in the history of life has not been constant, but that new places remain to be, not filled, but found and even created. And, strangely, that creation will be the accomplishment, at least in part, of the placeholders themselves acting and reacting on one another.[13] Thereby do progress and improvement

[13] Darwin's repeated use of the language of action and reaction (e.g., Chapter III, p. 75; Chapter VI, p. 178; Chapter XI, p. 350) immediately recalls Newton, as does his insistence on a gradualism which he understands to

remain possible, with absolute perfection lying eventually in view ("And as natural selection works solely by and for the good of each being, all corporeal and mental endowments will tend to progress towards perfection," he writes in the second-to-last paragraph of the abstract, Chapter XIV, p. 489). The very τελος that demands place as an end (one sense of τελος) to the signification of placeholders also demands, in order to retain the possibility of perfection (a meaning also related to τελος), the production of new places:

"Nor do I believe that any great physical change, as of climate, or any unusual degree of isolation to check immigration, is actually necessary *to produce new and unoccupied places* for natural selection to fill up by modifying and improving some of the varying inhabitants. For as all the inhabitants of each

---

proceed via "infinitesimally small inherited modifications" (e.g., Chapter IV, p. 95) and to require a summation over something approaching infinity (e.g., Chapter XIV, p. 481: "The mind cannot possibly grasp the full meaning of the term of a hundred million years; it cannot add up and perceive the full effects of many slight variations, accumulated during an almost infinite number of generations."). Yet he is hardly sanguine that a Newton of the grass blade awaits in the wings: "Throw up a handful of feathers, and all must fall to the ground according to definite laws; but how simple is this problem compared to the action and reaction of the innumerable plants and animals which have determined, in the course of centuries, the proportional numbers and kinds of trees now growing on the old Indian ruins!" (Chapter III, p. 75). At the same time, however, his pessimism by no means relinquishes a Newtonian project whose idiom he largely has appropriated (particularly in his strained recourse to laws, many of which are inarticulable); and one wonders if his attempt to ground an otherwise promiscuous representational relay isn't rooted in something like Newton's absolute space and time.

country are struggling together with nicely balanced forces, extremely slight modifications in the structure or habits of one inhabitant would often give it an advantage over others; and still further modifications of the same kind would often still further increase the advantage. No country can be named in which all the native inhabitants are now so perfectly adapted to each other and to the physical conditions under which they live, that none of them could anyhow be improved…" (Chapter IV, pp. 82-3; italics added).

And again in Chapter VI, p. 178:

"[N]atural selection can do nothing until favorable variations chance to occur, and until a place in the natural polity of the country can be better filled by some modification of some one or more of its inhabitants. And *such new places* will depend on slow changes of climate, or on the occasional immigration of new inhabitants and, probably, in a still more important degree, on some of the old inhabitants becoming slowly modified, with the new forms thus produced and the old ones acting and reacting on each other." (italics added)

Both quotations hover over an interesting ambiguity: are unoccupied places necessarily also new? Are new places necessarily also unoccupied? Darwin's answer is suggested by the example to which he returns to prove his point, the ready naturalization of 'foreigners': continuing where I left off above from Chapter IV, he writes, "…for in all countries, the natives have been so far conquered by naturalized productions, that they have allowed foreigners to take firm possession of the land. And as foreigners have thus everywhere beaten some of the natives, we may safely conclude that the natives might have been modified with advantage, so as to have better resisted such intruders" (Chapter IV, p. 83). Thus, unoccupied places may well be new in one place and occupied somewhere else; in a strange way, place itself has place and can be displaced; and his proof of this is precisely the replacement of placeholders by displaced placeholders. Unoccupied and new places, then, are still figured, paradoxically, by placeholders, ones who formerly occupied the place in a different place. On this explanation, 'new' places are both discovered (since they existed before) and created (since they existed before in a different place) by the arrival of new placeholders. Further, new places are also produced by placeholders "acting and reacting on each other." Place is thus both prior to the placeholder, though revealed by him, and consequent to the placeholder, though required by him. But, within this complicated deployment of "place," can it be clear anymore how Darwin would distinguish place from placeholder? Has not the metaphorical ground seemingly provided by the metaphor of place been revealed as anything but *terra firma*?

Yet it is ground that Darwin will not relinquish, even as his bind reaches its inexorable climax. The context is, not-incidentally, a discussion of his only diagram (reproduced here on page 133), the sort of diagram which he confesses elsewhere "it is almost impossible to do … without" when tracing the relationships between the living and the dead (Chapter XIII, p. 431), and which, as material substrate, he cannot help but continually reinterpret over the course of the book, as representing individuals, varieties, species, genera, families or orders, as well as thousands, tens of thousands or millions of generations, the geological strata of the earth, the unshown parts of the diagram and of course, with a similarly vertiginous recursivity, a tree.[14] A tree, an example of life, would also represent all life including itself, like that privileged placeholder, perhaps somehow imagined as emptied of all content, that would hold the place of place itself. In the context of this curious and almost indispensable substrate – the "almost" being as interesting as the near-impossibility – Darwin reveals shockingly that it is *place* according to which natural selection acts:

"[F]or natural selection will always act according to the nature of the places which are either unoccupied or not perfectly occupied by other beings; and this will depend on infinitely complex relations" (Chapter IV, p. 119).

Our confusion should be extreme. How does natural selection, helpless to do anything until chance variations arise, nonetheless act according to the nature of a place, especially a place that is unoccupied – either because empty or not yet there, if these really are distinguishable – or, again if this too is really distinguishable from the other cases, imperfectly occupied? How does such a place even have a nature? How is a place's nature available to Nature, in order for natural selection to act in accordance with it? And finally, if this acting depends

14 "...[O]ne species giving rise first to two or three varieties, these being slowly converted to species, which in their turn produce by equally slow steps other species, and so on, like the branching of a great tree from a single stem, till the group becomes large." Chapter X, p. 317.

on "infinitely complex relations" (we should emphasize, among placeholders!), how can Darwin remain certain that the number of places is not, at least potentially, indefinitely large – the thwarted ambition of one language and a name made for ourselves instead confused and scattered over the face of the earth? The fundamental problem, the problem of a fundament, is that place allows Darwin to preserve an enduring presence outside the endless relay of representation only on the condition that it never appears as itself, but, Darwin also requires, place must somehow interact with and through natural selection so that the proper and perfect placeholder can be found, even though such a placeholder would have to be indistinguishable from place or would itself be subject to replacement. But in that indistinction, would it not replace place, take the place of place, destroy place and be place? Voila evolution's end, a placeholder taking place's place.

We should ask, of course, with what impregnable defenses this placeholder-place would immure itself from invasion. Certainly it would entail the end of reproduction as both unnecessary for that which is at last fully present and, in the natural dynamic of parricide enforced by a finite world, as threatening.[15] It would entail as well the end of growth, of change and, ultimately, of motion:[16] for each of these, in the possibility of even temporary mismatch between the enduring place and its now enduring placeholder, unavoidably implies the condition of eventual replacement. This placeholder, no

[15] In other words, the end of deferral, not only unnecessary but incompatible with full presence – if there is such a thing.

[16] In short, the end of difference; and, taken with deferral's end, the end as well of *différance*.

longer *representing* a place but at last in its perfection *present* as place, motionless, changeless, ungrowing and unreproducing, if it is life at all, will be life entombed,[17] life interred,[18] almost like a fossil in its stratum or a written word, even a mere letter on a diagram, preserved somehow forever in a volume of history. But it will be the last fossil and the last word, not the last volume of history but history as the last volume, and its coffin will be itself.

Its correlate would be the perfect book promised by the abstract that is On the Origin of Species. Already noted is Darwin's admission that the abstract "must necessarily be imperfect." That necessity is linked to seemingly inexorable exigencies, namely a want of space and of time. Darwin's health has not been strong. Further, exacerbating that encroaching spatial and temporal finitude is the threatened competition of Wallace's strikingly similar "memoir." In potentially taking the place that Darwin imagines for his future work, it compels Darwin to send forth into the world a premature birth if only as a placeholder for the eventually mature and perfect work to come. In publishing it, however, Darwin does not give up the possibility of its eventual perfection; far from it. He merely defers that perfection – a perfection which will require difference from its current form – to a future date, should space and time be found. The question for us, however, is whether the perfection of such a work is really imaginable. For Darwin, on the other hand, this is less a question than an assumption that organizes his entire

---

17 Chapter XI, p. 311 ("Each word... may represent the apparently abruptly changed forms of life, entombed...").

18 Chapter XII, p. 332 ("If, for instance, the genera... were disinterred...").

thought. Thus, precisely in order to make some difficulties disappear, he concludes Chapter IX with a telling metaphor:

"For my part, following out Lyell's metaphor, I look at the natural geological record, as a history of the world *imperfectly kept*, and written in a changing dialect; of this history we possess the last volume alone, relating to only two or three countries. Of this volume, only here and there a short chapter has been preserved; and of each page, only here and there a few lines. Each word of the slowly-changing language, in which the history is supposed to be written, being more or less different in the interrupted succession of chapters, may represent the apparently changed forms of life, entombed in our consecutive, but widely separated, formations. On this view, the difficulties above discussed are greatly diminished, or even disappear" (pp. 310-1; italics added).

It neither occurs to Darwin whether history is truly imaginable on the condition of its being perfectly kept, nor that the interruptions marking what would otherwise be a perfect record could be anything other than accidental. He considers a history written in words, but not the words written in space – spaces which distinguish one word from the next and which, interpenetrating the word, distinguish one letter from the next too.[19] Indeed, specifically in order to

---

[19] And yet he also recognizes, at least with regard to the fossil history of life, that the story we tell depends on the gaps, however accidental; and that we would reach different conclusions with different gaps. Referring to his one and only diagram (reproduced here on p. 133), he writes, "By looking at the diagram we can see that if many of the extinct forms, supposed to be

demonstrate the non-essential role of extinction in his theory, to prove that it "has only separated groups; it has by no means made them," he goes so far as to imagine, and to consider as imaginable, the sudden reappearance of "every form which has ever lived on this earth" and concludes that, "though it would be quite impossible to give definitions by which each group could be distinguished… nevertheless a natural classification, or at least a natural arrangement, would be possible" (Chapter XIII, p. 432). But would it? Why does he still think there would be groups, though they could not be distinguished by definition? Is separating groups really incidental to making them? Isn't this why he has to qualify classification into arrangement – because categories from "species" to "kingdom" cease to be conceivable in this imagined world? But is an arrangement really any more conceivable? There are, of course, the incredibly concrete questions – *where* would this arrangement be? *when* would there be time for it? *how* would we apply common as opposed to proper nouns to each of its members? *why*, in a world of complete preservation, would it be necessary? And there is a no less concrete observation: the premise of the thought experiment is antithetical to Darwin's theory. In a world of complete preservation, there is no

---

embedded in the successive formations, were discovered at several points low down in the series, the three existing families on the uppermost line would be rendered less distinct from each other. If, for instance, the genera a^1, a^5, a^{10}, f^8, m^3, m^6, m^9, were disinterred, these three families would be so closely linked together that they probably would have to be united into one great family..." (Chapter X, p. 332). Darwin will acknowledge this dependence on accident without ever confronting accident's inexorable and unavoidable necessity. His ultimate history, whether of life or anything else, would be a history excised of all accident – as if any history would remain.

selection. Though throughout the book Darwin insists that "[t]he extinction of old forms is the *almost* inevitable consequence of the production of new forms" (Chapter X, p. 343), that extinction is "an *almost* inevitable contingency" (Chapter XIV, p. 471) and that "the extinction of species and of whole groups of species, which has played so conspicuous a part in the history of the organic world, *almost* inevitably follows on the principle of selection" (Chapter XIV, p. 475; italics added in all cases), he questions the necessity of these conspicuous absences to history no more than he does his own optimistic modification of inevitability with the adverb "almost."[20] The "almost" allows him to treat as contingent what is instead absolutely necessary – as soon as the world again becomes limited, when space becomes wanting and time short. In that world, our recourse must be to diagrams "almost impossible" to dispense with, in order to take the place of, and make meaningful, the tangled skein of infinitely complex relations, especially and explicitly between the living and the dead; and in that world, in parallel, extinction doesn't untie but cuts the Gordian knot of proliferating life, making the further perfection of life through selection possible on the condition of enforcing the imperfection of its record. Darwin at his most rigorous realizes this, whether in acknowledging that "we shall certainly never succeed in making so perfect a collection" [of "all the forms in any class which have lived throughout all time and space"] (Chapter XIII, pp. 432-3) or in confessing that his abstract "must necessarily be imperfect."

[20] This despite describing his own doctrine as that "of the extermination of an infinitude of connecting links..." (Chapter XIV, p. 463).

But he never takes the next step to examine the necessity of that imperfection, to consider how, in the context of spatial and temporal finitude, it is the expression of motion, of growth, of reproduction – in short of life – that both drives and is driven by the perpetual replacement and displacement of the placeholders he so assiduously tracks. A symptom of his failure to do so is his extremely convoluted discussion of the possibility of perfection. He doesn't believe in it. And he does. On the one hand, perfection unexpectedly comes in degrees and stages: there are organs in a "high stage of perfection" (Chapter VI, p. 183) and others in a "moderately high stage of perfection" (Chapter VI, p. 187), while natural selection "adapts the inhabitants of each country only in relation to the degree of perfection of their associates" (Chapter XIV, p. 472; see also, e.g., Chapter VII, p. 235 and Chapter XI, p. 379). He concludes that "there is no logical impossibility in the acquirement of any conceivable degree of perfection through natural selection" (Chapter VI, p. 204). On this reading, perfection is incomplete, relative and, on a rigorous interpretation of the word, accordingly imperfect. He admits that even "the most perfect organ, the eye" is imperfect for its manner of correcting the aberration of light (so it is said, anyway, "on high authority"; Chapter VI, p. 202) – though earlier its "extreme perfection" was a significant theoretical difficulty (at the start of the section entitled "Organs of extreme perfection and complication," he writes, "To suppose that the eye… could have been formed by natural selection, seems, I freely confess, absurd in the highest possible degree," Chapter VI, p. 186). At the end of Chapter VI, he even gives us some insight into what he could mean by a perfection that is incomplete or relative: "Natural selection in each well-stocked country, must act chiefly through the competition of the inhabitants one with another, and consequently will produce perfection, or strength in the battle for life, only according to the standard of that country" (p. 205).[21]

Yet, on the other hand, despite all of this, he refuses to relinquish the possibility of 'absolute perfection.' Even when he does. "Natural selection will not produce absolute perfection," he asserts unambiguously in Chapter VI (p. 202), before inexplicably continuing "nor do we always meet, as far as we can judge, with this high standard under nature." His use of "always" is perplexing: does he mean to suggest that at least some of the time we *can* expect to meet this high standard under nature, although natural selection will not produce it? A few pages later, he introduces a further important qualification: "Natural selection will not *necessarily* produce absolute perfection; nor, as far as we can judge by our limited faculties, can absolute perfection be everywhere found" (p. 206; italics added). "So again," he writes on p. 211, "in some few cases, certain instincts cannot be considered as absolutely perfect..." By implication, then, absolute perfection is not merely a regulative idea; Darwin appears to believe in its real possibility, in its discovery, if not everywhere, then at least here

[21] He reinforces this understanding of perfection in Chapter XI, p. 379: "I suspect that this preponderant migration from north to south is due to the greater extent of land in the north, and to the northern forms having existed in their own homes in greater numbers, and having consequently been advanced through natural selection and competition to a higher stage of perfection or dominating power, than the southern forms."

and there.[22] While before we marveled because "the belief that an organ so perfect as the eye could have been formed by natural selection, is more than enough to stagger any one" (p. 204), by the end of the book (Chapter XIV, p. 472), "[t]he wonder indeed is, on the theory of natural selection, that more cases of the want of absolute perfection have not been observed." But how would one observe absolute perfection or its want? If by "perfection" Darwin means "strength in the battle for life," is absolute perfection simply unvanquishable strength? Anything less than that is of course not absolute. But to see no difficulty in distinguishing the unvanquishable from the merely unvanquished could only be a sign of blindness.

In short, Darwin's book forever arrives at solutions that only repeat the problems they would solve. The individual being, whose transitoriness is at once the inevitable outcome of Darwin's account and its necessary condition, must accordingly represent other beings, past (ancestors), present (conspecifics) and future (descendants). But those other beings are no less transitory. Hence they instead represent, however imperfectly, a place; that would be their foundation, the enduring presence that they themselves strive, and fail, to be.

---

22 Indeed, when discussing the cell-making instinct of the hive-bee, he declares that "Beyond this stage of perfection in architecture, natural selection could not lead; for the comb of the hive-bee, as far as we can see, is absolutely perfect in economizing wax" (Chapter VII, p. 235). He can't help but qualify his own declaration, however, in two ways – first, by acknowledging the limitations of our vision; and second, by limiting absolute perfection to a particular realm, that of economizing wax. Should the acuity of our vision increase, or should natural selection favor some other economy than wax, for example, what will have become of this absolute perfection?

That failure of perfect representation then becomes the driving force in natural selection, which, Darwin thinks, is ever favoring forms better representing their place. His proof of this, however, can never detach itself from the placeholders that alone are available for selection in place's stead, until finally he mobilizes place itself, displaces it, and in thereby endeavoring to release it from the veil of its placeholder, reveals it as itself a placeholder. Place, now come into play as unoccupied or new, thus destabilizes the possibility of eventual and absolute perfection in the very act of guaranteeing it. That is to say, the end promised by perfection, both as the stopping point of this representational regress and as τελος, is preserved as a possibility only by deferring it, by recognizing as a placeholder what had hitherto held the place of that end. Put otherwise, perfection is conceivable to Darwin in part on the condition, inevitably treated as accidental rather than indispensable, of forever demonstrating its absence. Darwin is, on the one hand, always doing this (and his theory is unimaginable without it) and, on the other, always forgetting it. He can and cannot see the want of absolute perfection. Were he to see it clearly, the nature of this being which Nature tends, and the meaning of "being," would necessarily emerge as a profound question – which it does not, for him. Were he not to see it clearly, the nature of a perfect being, indistinguishable from the place that it represents but also in its inert eternality unidentifiable as life, would emerge as a profound problem – which it does not, for him. He can only hover indeterminately between these two options, dimly seeing what is and is not there.

The mirror image of this pursuit of perfection, of an end, is the parallel pursuit of an origin. Darwin never abandons this notion, even as he forever defers it, the origin always retreating to an earlier horizon. Breeds lack distinct origins and species originate from other species. His only diagram in the book (reproduced on the next page) is a concise representation of this representational problem. Its structure is recursive and algorithmic. Darwin uses its horizontal axis to represent every taxonomic category from individuals to orders, while he has its vertical axis represent years and generations of any numerical extent, as well as geological strata. The parallel horizontal lines stretching across the diagram represent moments when we happen to be looking; but the other lines, stretching from ancestor to descendant, remain visible even though we weren't looking. If we understand the diagram as a tree, it is with the concession that trees could also be represented on it. Similarly, the broken lines that extend downward from the capital letters at the base promise that this diagram also represents the unshown portions of itself, the branching ramifications which culminate in the provisional capital-letter origins from which we started. For, "[w]e will suppose the letters A to L to represent allied genera, which lived during the Silurian epoch, and these have descended from a species which existed at an unknown anterior period…" he writes in Chapter XIII, p. 420. Even the beginnings, then, are descendants.

Yet Darwin will no more relinquish the possibility of an eventual, original origin than the corresponding possibility of eventual, final perfection – even though, and in a certain way even because, he cannot find it. He criticizes some opponents for the belief that innumerable times "certain elemental atoms

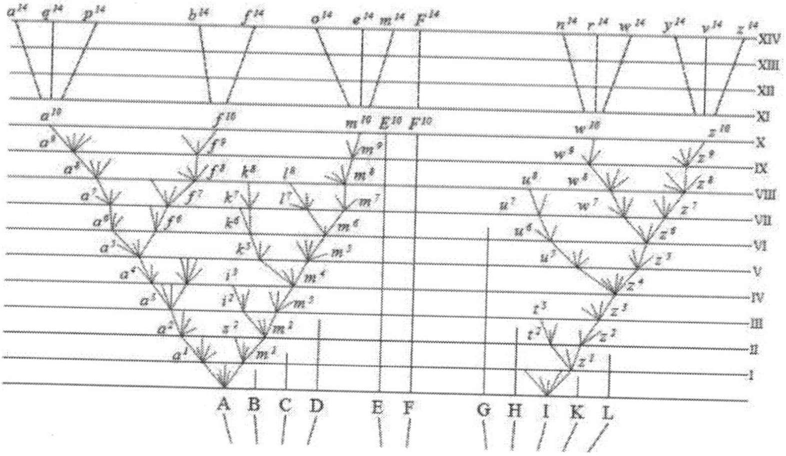

have been commanded suddenly to flash into living tissues" and argues that such naturalists thereby "ignore the whole subject of the first appearance of species" (Chapter XIV, p. 483). But can Darwin himself honestly claim to have treated of this subject? Is a breath a more suitable answer than a flash? "I believe that animals have descended from at most only four or five progenitors, and plants from an equal or lesser number," he writes, before continuing:

"Analogy would lead me one step further, namely to the belief that all animals and plants have descended from some one prototype. But analogy may be a deceitful guide. Nevertheless all living things have much in common… Therefore I should infer from analogy that probably all the organic beings which have ever lived on this earth have descended from some one primordial form, into which life was first breathed." (Chapter XIV, p. 484)

Does analogy really lead to this step? Analogy to what? The analogies of one living being to another? Assuming *concesso non dato* that the analogy is truly a characteristic of the entities compared and not in some way the contribution of the comparer, Darwin himself has been at pains to discredit this deceitful guide, for example, in Chapter XIII (p. 427): "... analogical or adaptive character, although of the utmost importance to the welfare of the being, are almost valueless to the systematist. For animals belonging to two most distinct lines of descent, may readily become adapted to similar conditions, and thus assume a close external resemblance..."[23] Analogy of this sort by no means leads inexorably to the conclusion of a single origin.

Nor, in general, does Darwin's argument, as can be succinctly demonstrated from his one and only diagram. Let us leave aside for now the significant problem of what a kind is – to be returned to in a moment – and instead re-examine that diagram. While it appears to show a progressive increase in number of kinds, from an initial 11, signified by capital letters at the bottom, to an eventual 15, signified by lowercase letters at the top – and further, by its slanting, broken lines beneath the capital letters, suggests that the 'original' 11 are themselves

---

23 This resemblance Darwin is at pains to attribute, not to "place" in the rather unusual sense in which he deploys that term, but rather to "zones": "[W]hen varieties enter any zone, they occasionally assume some of the characters of the species proper to that zone" (Chapter XIV, p. 472). But what do zones enable that place does not?

descendants, that there were fewer kinds at an even earlier stage and that, if extrapolated far enough, perhaps a single kind would lie at the absolute origin – yet nothing in his argument necessitates such increases of kinds over time. Darwin knows very well that the geological record, for example, is far more complicated, and indeed that some periods stand out as especially creative and others as especially destructive. In fact, his own diagram, on closer inspection, is also more complicated: if we had stopped our analysis at the horizontal line labeled IX or X, rather than the one labeled XIV (signifying, for example, the nine-thousandth, ten-thousandth or fourteen-thousandth generation), we would have seen with the advance of time a contraction rather than expansion of kinds. During that period, capital letter A gave rise to three kinds, I to two, E and F to one apiece, and B, C, D, G, H, K and L to none – a net loss, if all kinds are assumed to represent the same taxonomic category, of four. As the diagram demonstrates, the idea that there were fewer kinds in the past, and eventually there was but a single kind, ancestor not only to everything that exists now (conceivably the descendants of an ancestor whose neighbors left no progeny) but to "*all* the organic beings which have ever lived on this earth," is therefore not a conclusion, analogical or otherwise, from Darwin's argument, but at best an unnecessary assumption. Unnecessary because, in his actual description of the origins of varieties, species, genera, etc., Darwin never invokes it. Far from it; at every provisional origin that he describes, we instead encounter a community of beings whose infinitely complex mutual relations define one another. Characteristically, Darwin both works out the implications and fails to see them.

To these we must now turn.

III. The tropology of essential relations

The imperfect presence of every organic being – its incompleteness, its presence-pervaded-by-absence – forces Darwin into a referential structure, where that being provisionally and only partially takes the place of its ancestors and holds the place of its descendants. For Darwin, that structure is explicitly metaphorical and is governed by likeness, though he cannot help but problematize that, as when likeness of an analogical or adaptive character is instead "a deceitful guide." As much as inheritance would be the rule by which Darwin regulated biological rhetoric, he is compelled to admit exceptions. Thus, corresponding to and being confused with this seemingly internal potency passed from parent to offspring is an external and circumstantial one; adaptation to similar conditions, rather than common descent, can also drive perceived resemblance.[24] As with so many of the problems that Darwin encounters, this one too, when cast in a different light, will be seen by him as a solution – and, of course, vice versa. For, as always, the addition of a supplement which happily completes the presence of the organic being is, from a different perspective, precisely what also disrupts that presence. Hence

---

[24] Thus, despite "propinquity of descent [being] the only known cause of the similarity of organic beings" (Chapter XIII, p. 413), Darwin has to acknowledge, if only to dismiss, similarity due to external causes. He might instead have wondered if the conditions of life, as he called them, weren't also a variety of inheritance.

the problematic rescue, by way of place, which we have just finished examining, one which, for Darwin, preserves the possibility of eventual perfection, of a presence that is at last fully present, only at the necessary expense of extinguishing – by mercilessly exposing the limitations of its provisional presence – each successive placeholder in turn.

The same dynamic is about to play out again, and again it will drive Darwin to rhetoric. Revealing the incomplete presence of the organic being, it will be the rhetorical motor that propels that being to the full presence which is its birth-right; yet, failing to stop what it succeeded in starting, it will accompany that being every step of the way, propelling it thereby towards an always receding horizon. A finite trajectory will thus be disclosed as an orbit. The new player in this drama is the other half, the first half and, he says, the less important half, of the "large and metaphorical sense" in which Darwin uses the term 'struggle for existence.' As already quoted: "I should premise that I use the term Struggle for Existence in a large and metaphorical sense, including dependence of one being on another, and including (which is more important) not only the life of the individual, but success in leaving progeny" (Chapter III, p. 62).

We have not sufficiently paused over how either of these understandings is metaphorical, or how that metaphor applies, not as one might expect, to the organic being – at least not in any simple sense – but to the struggle. Etymologically understood, a metaphor is a carrying (φερω) across, after, over or with (μετα); it is, almost literally, a translation in the several senses of that word (whose Latinate roots mirror the Greek ones, albeit via a perfect passive participle [*latus* < *fero*, cognate of φερω] which makes of translation an accomplished metaphor, a "having been carried across"). In metaphor, this carrying across is initiated performatively, by decree, through the paradoxical gesture of declaring two different things the same. Hence the gulf being crossed is the difference between beings. Yet, the traditional rationalization of the paradox – the invocation of a shared but hidden commonality, even essence[25] – complicates this: is it the gulf among beings, or is it the gulf within being, which metaphor would span?[26] As we shall see, this question is not incidental to the two understandings that Darwin proposes for the term 'struggle for existence.' For, the differences among beings and within being are the conditions of struggle, selection and change. In the first metaphorical sense adumbrated by Darwin, the gulf appears among beings, and it is this which shapes the interrelationships (including dependence) by which, however provisionally, those beings surmount the limitations that define them and that announce themselves as differences. But in his second metaphorical sense, it is within being: out of those differences would emerge the perfected being in whom its plural precursors, though annihilated, find their perpetuity and completion and of whom they were always a partial expression.

---

25 The manner in which a shared commonality is aggrandized into an essence will be discussed later, specifically in the context of a passage from Plato pertaining to the definition of biological kinds.

26 That is, if the two beings being equated despite their apparent separateness in fact share a commonality or even essence, how is it that they nonetheless became separated in the first place? It is this dispersal of commonality within difference, or this segregation of commonality by difference, which might be read as a gulf within being.

But let's work this out more slowly. How are we to grasp the struggle for existence as metaphorical? In the context of the earlier discussion, the second of the two aspects that Darwin highlights seems clearer than the first and supports my analysis thus far. "Success in leaving progeny" points to the diachronic recuperation by which Darwin preserves the possibility of presence for a being which, after all, is struggling to exist – or, to put it more strongly, is struggling to come into (full) existence. That struggle of course qualifies how 'being' is to be understood; and it is a sign of Darwin's perhaps unwitting rigor that he applies the metaphor, not to that being – any example of which, if the meaning of being is understood as presence, has yet quite to be – but to the struggle. Darwin's convoluted discussion of perfection, however, suggests that he himself is unclear whether the struggle has an end; his descriptions depict an indefinite *perfectibilité* that his declarations of absolute perfection contradict.[27] A sign of Darwin's ultimate commitment to absolute perfection and to the metaphysics of presence which is its condition, however, is nothing other than his presumption that his use of the term 'struggle for existence' is metaphorical rather than, say,

[27] I choose the word "*perfectibilité*" in part to allude to Rousseau's A Discourse on Inequality, where *perfectibilité* (often translated as a faculty of "self-improvement" or simply "perfectibility") enables human beings both to develop their other faculties and to undo them, so that human beings rise above beasts and fall beneath them. Rousseau later in Part One specifically associated *perfectibilité* with language. Although it is beyond the scope of the present argument, I understand *perfectibilité* in Rousseau's text as pointing to and following from the indefinite rhetorical potential of language. The manner in which the predicament of being, in turn, is also linguistic – and even more specifically, graphic – will emerge over the course of this chapter.

catachrestic. His metaphor can be rationalized; it is anchored to the stability of a shared commonality. Thereby in principle the diachronic recuperation can eventually come to rest on synchronic grounds – specifically, for Darwin, place. However tortuous, his deployment of place would thus underwrite a great accomplishment (indistinguishable from an expression of overwhelming desire): the condensation of a struggle among beings – including Oedipus-like offspring who exterminate their parents – into a struggle within being. Hence, progeny will be what their parents were; and it would be "preposterous," Darwin thinks, "if a perfect kangaroo were seen to come out of the womb of a bear" (Chapter XIII, p. 425) – though earlier he saw no difficulty in a race of bears gradually becoming very like a whale. In short, metaphor's appeal is its power to efface the difference that is its condition; but its danger is an effacement without end: bears which become not only whales but kangaroos, places which become placeholders, and metaphors which become – if they weren't always already – catachrestic.

What about the first half of Darwin's large and metaphorical sense? How is dependence of one being on another to be understood as metaphor? On the face of it, the structure of such dependence on the contrary seems metonymical. For there is nothing which suggests that the beings depending on one another are related by any intrinsic commonality (such as heredity implies for Darwin) – much less by a shared essence. Rather, dependence, like metonymy, opens up relationships of happenstance: after all, there is nothing about a street named Wall that essentially connects it to a global financial system; it is merely that many of the system's players happen, for complex historical and contextual reasons, to have offices on that street. Likewise, the dependence of one being on another expresses only what might playfully be called the essential inadequacy of each of those beings, an inadequacy overcome, as it were, contingently through the accident of available, complementary beings. Only if such a lack can be an essence will we have found a hidden commonality that justifies the language of metaphor rather than that of metonymy. Even then, it will be on the condition of recognizing only a general, rather than specific, lack: for the interdependence of beings requires that they mutually supply each other's lacks.

The possible confusion of metaphor and metonymy thus signals a far deeper problem. The opposition between them is no more stable than that between essential and accidental, necessary and contingent, intrinsic and extrinsic. For what is there to ground the stability of those oppositions? If the metaphorical understanding of the seemingly metonymic dependencies is through the recognition of a shared commonality or even essence that, remarkably, is an absence[28] – and not even the same absence – can we be certain that this is really different from the apparently more secure metaphorical understanding of success in leaving progeny? As argued above, Darwin is rigorously driven to understand his metaphorical sense as applying, not to organic beings, but to the struggle for existence, precisely because no organic being endures long enough to latch on to – none is sufficiently present to anchor a metaphor indefinitely. Nor, despite valiant attempts, does

[28] Hence, the gap between beings would be bridged only by displacing that gap within being itself – where it is by no means less threatening.

Darwin's recourse to place provide such a grounding. Yet Darwin never questions the eventual appropriateness of the term metaphor to the same degree that he never questions the eventual triumph of presence. Everything he describes, however, calls into question that triumph and, in parallel, threatens to expose his metaphor as secretly catachrestic, as relying not on the eventual closure of his system so much as on its perpetual openness – its incompleteness, imperfection, blanks, gaps or absences. If the previous paragraph suggested a way to read Darwin's seeming metonymy as metaphorical, here we begin to approach something like its converse, a reading of his seeming metaphor as metonymical – where a presumed commonality or even essence turns indistinguishably into a circumstantial proximity, and "merely" analogical or adaptive resemblance is mistaken for hereditary likeness. An internal sameness may yet be revealed as an external imposition.

Indeed it is precisely the permeability characterizing the border between inside and outside, metaphor and metonymy, success in leaving progeny and dependence of one being on another, which breaches the absolute limit of perfection. Writing of the diversity of appendages found in various birds, Darwin notes, "Yet the structure of each of these birds is good for it, under the conditions of life to which it is exposed, for each has to live by a struggle; but it is not necessarily the best possible under all possible conditions" (Chapter VI, p. 182). Hardly surprising; after all, how could it be? Natural selection only ever operates under specific conditions, not all possible ones. The specificity of that operation necessarily stymies the discovery of a perfection whose absoluteness would deliver it from conditionality.[29] And contributing to that specificity is not simply the gamut of all possible physical conditions; it is the mutual relations of all organic beings with each other. Indeed, these mutual relations are typically, on Darwin's view, far more important than the physical conditions – as he frequently acknowledges, nowhere more emphatically than in his discussion of "the deeply-seated error," his prioritizing of physical over biotic relations, that for a while constrained his appreciation of the Galápagos (Chapter XII, p. 400).[30] These mutual relations cast perfection always toward the horizon because, as Darwin repeatedly stresses, they are both themselves in play and "infinitely complex." To wit:

"Owing to this struggle for life, any variation, however slight and from whatever cause proceeding, if it be in any degree profitable to an individual of any species, in its *infinitely complex* relations to other organic beings and to external nature, will tend to the preservation of that individual, and

---

[29] Hence, as earlier observed in footnote 22, the one example of absolute perfection declared by Darwin is immediately qualified: "the comb of the hive-bee... is absolutely perfect *in economizing wax*."

[30] See also, e.g., Introduction, p. 6: "Yet these relations [the mutual relations of all the beings which live around us] are of the highest importance, for they determine the present welfare and, as I believe, the future success and modification of every inhabitant of this world"; Chapter XI, p. 350: "…the relation of organism to organism being… the most important of all relations…"; Chapter XIV, p. 477: "...the relation of organism to organism is the most important of all relations..."; Chapter XIV, p. 487: "...[A]nd as the most important of all causes of organic change is one which is almost independent of altered and perhaps suddenly altered physical conditions, namely, the mutual relation of organism to organism..."

will generally be inherited by its offspring." (Chapter III, p. 61; italics added)

"Let it be borne in mind how *infinitely complex* and close-fitting are the mutual relations of all organic beings to each other and to their physical conditions of life." (Chapter IV, p. 80; italics added)

Under these infinitely complex circumstances, is it surprising that a structure is "not necessarily the best possible under all possible conditions"? Indeed, what prospect, concern or desire justifies Darwin's use of the adverb "necessarily," thereby leaving open the possibility of, not just an eventual, but even a present absolute perfection? Under what conditions can a structure or organism adapt to all possible conditions? The potential that an adverb preserves Darwin's own description denies; for what he describes is more akin to Rousseau's inexhaustible *perfectibilité* than to an achievable, much less achieved, perfection. Infinitely complex mutual relations lead to similarly complex co-adaptations, which in turn alter the mutual relations and, as a result, themselves alter. This almost recursive or algorithmic process can only thrust perfection, in the absolute sense, out of view, as Darwin sometimes does and does not see:

"...I can see no limit to the amount of change, to the beauty and infinite complexity of the coadaptations between all organic beings, one with another and with their physical conditions of life, which may be effected in the long course of time by nature's power of selection." (Chapter IV, p. 109)

Though seeing no limit, he will also insist on those limits, whether in premising an eventual perfection or simply in identifying as if it were stable and inherent the function of an organ. For every characteristic has a meaning (e.g., an organ's function or an organism's ecological role) which is inextricably bound by context – but the possible contexts themselves are without bounds. Hence perfection, which would be an absolute limit, is itself without limit. Perfection is incomplete, indefinite and, therefore, imperfect. The already discussed disruption of the desired stability of place is simply another form of this problem: if, as quoted before, "natural selection will always act according to the nature of the places which are either unoccupied or not perfectly occupied by other beings," Darwin nonetheless can't help but observe, "this will depend on infinitely complex relations" (Chapter IV, p. 119). Place remains contextual and thus subject to, as it were, constant reinterpretation – which can only lead to the confusion of place and placeholder already described. Absent the stability of place, however, the second metaphorical sense of the struggle for existence becomes ungrounded. Darwin's professed belief in progress ("And as natural selection works solely by and for the good of each being, all corporeal and mental endowments will tend to progress towards perfection"; Chapter XIV, p. 489) is likewise undone or at least significantly qualified: progress will be, like everything else, conditional – and the improvement of corporeal or mental endowments will by no means imply their absolute superiority over prior forms,[31] only

[31] Darwin nonetheless suggests otherwise (Chapter X, p. 337): "If under a nearly similar climate, the eocene inhabitants of one quarter of the world

their better fit to an always provisional context which can shape and be shaped by them.

For Darwin it is indeed a question of such shaping. Here a second aspect of the permeability characterizing the border between inside and outside becomes legible and with it the way in which his two metaphorical senses of the term 'struggle for existence' interpenetrate and mutually imply one another. Not only are organic beings "most remote in the scale of nature… bound together by a web of complex relations," an understanding of the action and reaction of which renders the physicist's calculation of the trajectories of a handful of falling feathers trivial by comparison (Chapter III, p. 75), but they literally *and essentially* shape each other:

"…[T]he structure of every organic being is related, *in the most essential and yet hidden manner*, to that of all other organic beings with which it comes into competition for food or residence, or from which it has to escape, or on which it preys. This is obvious in the structure of the teeth and talons of the

---

were put into competition with the existing inhabitants of the same or some other quarter, the eocene fauna or flora would certainly be beaten and exterminated; as would a secondary fauna by an eocene, and a palaeozoic fauna by a secondary fauna." Darwin is of course hedging his bets by pitting one geological era's inhabitants against those that closely followed it, and ostensibly by controlling for context by insisting on similar climates. But in the process has he not forgotten his own repeated insistence on the *infinite* complexity of the mutual relations among organisms? Might not slight differences, whether in climate or in species assemblage or even in temporal sequence – should this organism appear before that organism, for example – be amplified by that infinite complexity? Can such amplification simply be ruled out, thereby justifying Darwin's confident "certainly"?

tiger; and in that of the legs and claws of the parasite which clings to the hair on the tiger's body." (Chapter III, p. 77, italics added).

And again:

"...[W]hen a plant or animal is placed in a new country amongst new competitors, though the climate may be exactly the same as in its former home, yet the conditions of its life will generally be changed *in the most essential manner*. If we wished to increase its average numbers in its new home, we should have to modify it in a different way to what we should have done in its native country..." (Chapter III, p. 78, italics added)

The point is that an organism may be *essentially* shaped by something that is outside of it, a context into which it is born or thrown. Many of Darwin's inheritors will rapidly forget this possibility, all the more after the discovery of DNA, and will as a result confuse Darwin's "strong principle of inheritance" (e.g., Chapter IV, p. 127; Chapter XIII, p. 438) with instead a material and endogenous principle of heredity. But Darwin's own thinking about this is more pluralistic and subtle. "Owing to this struggle for life, any variation, however slight and *from whatever cause proceeding*... will generally be inherited by its offspring" (Chapter III, p. 61; italics added). Among the causes of variation that Darwin imagines are use and disuse, acclimatization or habit, and external conditions – though, if the latter is limited to climate or food, he considers its influence slight (Chapter V, e.g. p. 132). Note that all of these will potentially reflect an organic being's interactions with other such beings. Hence the causes of heritable variation are not restricted by Darwin to a merely internal potency. But it would not matter even if they were so restricted, because the (always provisional) *meaning* of that internal potency only emerges in the context of other organic beings. Inheritance, for Darwin, is always paired with context, a claim easily supported by restoring and differently inflecting what was hidden in the ellipsis of the quotation above: "Owing to this struggle for life, any variation, however slight and from whatever cause proceeding, if it be in any degree profitable to an individual of any species, *in its infinitely complex relations to other organic beings and to external nature*, will tend to the preservation of that individual, and will generally be inherited by its offspring." Variations generally inherited only have relational meaning. Thence "the many striking and complex coadaptations of structure between one organic being and another, which we see everywhere throughout nature" (Chapter V, p. 132). If these coadaptations have as their source, on a simplified scheme of some of Darwin's heirs, a mutation in DNA, that mutation is nonetheless without value except in terms of what it enables or prohibits in the struggle for existence with, whether as enemy or companion, other beings. Accordingly organisms, as much as they are, in modern parlance, the expression of their genes, are equally the expression of their neighbors, even far distant ones – and if they inherit the former, it is on the condition of having inherited a world along with them. And so organisms shaped by their parental inheritance – even from parents who were notably shaped differently – are also shaped by those with whom they interact, those in the contexts of whom their parental inheritance takes on (provisional) meaning. And these contexts, in a certain sense, are also inherited – that is, they are received as an irrefusable bequest – even while no boundaries can be staked for them.

A final version of the interpenetration of these two inheritances, as it were an internal (metaphorical) and an external (metonymic) one, respectively corresponding to the two metaphorical senses of the term 'struggle for existence,' is brought to light by Darwin's insightful blindness about the filiation of languages, which he uses to illustrate his understanding of classification. For, he insists, "[i]f we possessed a perfect pedigree of mankind, a genealogical arrangement of the races of man would afford the best classification of the various languages now spoken...The various degrees of difference in the languages from the same stock, would have to be expressed by groups subordinate to groups; but the proper or even only possible arrangement would still be genealogical; and this would be strictly natural... and would give the filiation and origin of each tongue" (Chapter XIII, pp. 422-3). Such a cavalier pronouncement in the linguistic realm has an equally cavalier correlate in the realm of natural history. There the usually cautious Darwin does not hesitate to declare that a plant's ability to graft onto a neighbor, whether closely related or not, is "entirely unimportant for its welfare in a state of nature" (Chapter VIII, p. 261). And yet plants graft onto each other all the time, as the preponderance of 'kissing trees' in many a forest testifies, and there is no *a priori* reason, other than a metaphysical prejudice, to exclude this common phenomenon as unimportant. Recent work in fact suggests otherwise: such grafts can lead, basically overnight, to new species and may also help to explain the mysterious prevalence of allopolyploidy among plants (see, for example, Fuentes et al., 2014). Likewise the genealogical arrangement of languages will always be stymied by a linguistic kiss, for example, the way that English, including Darwin's, became entwined with French after the Norman conquest, not to mention the proliferation of Latin, Greek or Latin-Greek hybrids throughout Darwin's ostensibly English text, specifically for naming organisms.[32] In sum: the borders separating languages

---

[32] Darwin is not inattentive to the import and difficulties of names. In Chapter I, an organism's history relies on its being named (p. 40: "But as yet they [improved individuals] will hardly have a distinct name, and from being only slightly valued, their history will be disregarded") and that history is not least an account of its naming (p. 40 continued: "When further improved by the same slow and gradual process, they will spread more widely, and will get recognized as something distinct and valuable, and will then probably first receive a provincial name"). Community of descent (that is to say, likeness or metaphor) will justify organic beings sharing, for example, a species name; while a metonymic relationship of parts one to another will justify the naming of homologous structures: "[P]arts may change to almost any extent in form and size, and yet they always remain connected together in the same order. We never find, for instance, the bones of the arm and forearm, or of the thigh and leg, transposed. Hence the same names can be given to the homologous bones in widely different animals" (Chapter XIII, p. 434). Indeed Darwin foresaw that "considerable revolution in natural history" which would release systematists from the vain pursuit of the essence of the term species, but only by delivering them to "a far more essential consideration": "whether the differences [between forms] be sufficiently important to deserve a specific name" (Chapter XIV, pp. 484-5). That the difficulties of such deliverance at times impressed themselves upon Darwin is clear from an October 2, 1852 letter to Joseph Hooker: "Are you

are anything but secure, and so too the borders separating organisms, in both cases regardless of their genealogic lineage – the concept of which is accordingly cast into confusion. In biology, it wasn't until the late 20th century that this phenomenon received serious attention with the discovery of genetic exchange between neighbors regardless of their (classically understood but now deeply problematized) relatedness. But the problem was, literally, already *legible* in Darwin's text, in its polyglot admixture of languages, and in his two metaphorical senses, neither of which he can rigorously keep apart.

Yet, if he cannot absolutely separate them, Darwin nonetheless wants to privilege one – success in leaving progeny – as more important than the other, a privilege that many of his heirs exaggerated monstrously. The attraction of that privilege, however fantastical, is understandable: for it would preserve an essential possibility. Against the backdrop of those infinitely complex mutual relations whose contingency challenges the possibility of an essence, this fantasy holds out the promise of an essence-restoring resolution. Thus the widespread view, whether in Darwin's time or our own, that

---

good at inventing names: I have a quite new & curious genus of Barnacle, which I want to name, & how to invent a name completely puzzles me" (as quoted in Stott, p. 82). As it should: does not his theory in a certain way both rely on and endeavor to explain the invention of names? (Has not genealogy always referred the bloodline to the proper name, and the proper name to the blood-line?) Would not a principal accomplishment of The Origin be a theory or rule for naming? One, moreover, which, through its own linguistic hybridity, confuses the very distinctions upon which it insists? Those who understand Darwin as "merely" a nominalist fail to perceive the anything-but-mere stakes of a name.

"there must be some *essential* distinction between species and varieties" (Chapter VIII, p. 268, italics added) and that such a distinction will be based upon the second metaphorical sense of the term 'struggle for existence,' i.e., reproductive success: "The fertility of varieties... when intercrossed, and likewise the fertility of their mongrel offspring, is, on my theory, of equal importance with the sterility of species [when intercrossed]; for it seems to make a broad and clear distinction between varieties and species" (Chapter VIII, p. 246). But Darwin, unlike many of his later students, knows that this can't work. It is undone by the fact that fertility and sterility themselves come in graduated degrees: "... it is most difficult to say where perfect fertility ends and sterility begins... It can thus be shown that neither sterility nor fertility affords any clear distinction between species and varieties..." (Chapter VIII, p. 248). In fact it is undone even before that, and along with it Darwin's assumption of a default Malthusian economy in the state of nature ("There is no exception to the rule that every organic being naturally increases at so high a rate, that if not destroyed, the earth would soon be covered by the progeny of a single pair" [Chapter III, p. 64]). For, Darwin well knows, there *is* an exception, which, all else failing, Darwin will use definitionally to *identify* unnatural states;[33] the exception is the

[33] See, for example, Chapter VIII p. 251: "Although the plants in these experiments appeared perfectly healthy, and although both the ovules and pollen of the same flower were perfectly good with respect to other species [i.e., in terms of successful fertilization], yet as they were functionally imperfect in their mutual self-action, we must infer that the plants were in an unnatural state." Later in the chapter (p. 277), he continues: "The sterility of hybrids, which have their reproductive systems imperfect, and which have had this system and their whole organization disturbed by being

extreme sensitivity of reproduction (the second metaphorical sense) precisely to context (the first metaphorical sense), the "[reproductive] system appearing to be far more susceptible than any other part of the organization, to the actions of any change in the conditions of life" (Chapter I, p. 8; see also, e.g., Chapter VIII, p. 273). As much as Darwin himself wishes to recover an absolute boundary between inside and outside which would perfectly insulate essences from the happenstance of relations, and success in leaving progeny from dependence of one being on another – the most important of the conditions of life – he is compelled (as his inheritors should have been compelled) to acknowledge the permeability of these borders.

What, then, are the implications of an essence infiltrated by relations? Or rather, does the irreducible importance of relations banish the possibility of an essence? Darwin never explicitly confronts these questions, while implicitly he answers them contradictorily. On the one hand, he continues

---

compounded of two distinct species, seems closely allied to that sterility which so frequently affects pure species, when their natural conditions of life have been disturbed." Yet he is well aware of examples in which hybrids are far from sterile, as in the case he cites from Herbert of "a hybrid from *Calceolaria integrifolia* and *plantaginea*, species most widely dissimilar in general habit, [which, according to Herbert] 'reproduced itself as perfectly as if it had been a natural species from the mountains of Chile'" (p. 251). He is also well aware that in some cases, the disturbance of "the natural conditions of life" may lead to excessive fertility – as with rabbits which "breed most freely under the most unnatural conditions" (Chapter I, p. 9). In this context, it is also curious to note that Darwin's proof of the tendency to geometric increase, so critical to his theory, is precisely the proliferation of "naturalized" productions – i.e., organisms introduced to new areas typically by man. See Chapter III, pp. 64-5.

to use the language of essences (as when he distinguishes adaptive from essential characters or characterizes a fauna or flora as essentially the same as another; for examples, see Chapter XI p. 349 and Chapter XIII p. 413) and to invoke related concepts, not least (as already noted) by keeping in reserve the prospect, if not the eventuality, of (absolute) perfection. This same Darwin looks at a world of myriad organic forms and is far from befuddled. It is he who asks in Chapter VI (p. 171), "Why is not all nature in confusion instead of the species being, as we see them, well defined?" It is also he who wishes to privilege as more important the second metaphorical sense of the term 'struggle for existence.' On the other hand, however, another Darwin emerges in a different context. He goes as a young man to the Galápagos Archipelago and is struck for the first time "how entirely vague and arbitrary is the distinction between species and varieties" (Chapter II, p. 48). Not coincidentally, the Galápagos also hosted his revelation that privileging physical over biotic conditions as more important to organisms is a "deeply-seated error" (Chapter XII, p. 400). This second Darwin sees a continuum where the first one sees a category: "Certainly no clear line of demarcation has as yet been drawn between species and sub-species... or, again, between sub-species and well-marked varieties, or between lesser varieties and individual differences. These differences blend into each other in an insensible series; and a series impresses the mind with the idea of an actual passage" (Chapter II, p. 51). It will be this Darwin who explicitly and repeatedly gives up the pursuit of essences. Two dramatic examples:

"From these remarks it will be seen that I look at the term species, as one arbitrarily given for the sake of convenience to a set of individuals closely resembling each other, and that it does not essentially differ from the term variety, which is given to less distinct and more fluctuating forms. The term variety, again, in comparison with mere individual differences, is also applied arbitrarily, and for mere convenience sake." (Chapter II, p. 52)

And again, most forcefully, in the final chapter as he prophesies "a considerable revolution in natural history" (pp. 484-5):

"In short, we shall have to treat species in the same manner as those naturalists treat genera, who admit that genera are merely artificial combinations made for convenience. This may not be a cheering prospect; but we shall at least be freed from the vain search for the undiscovered and undiscoverable essence of the term species."

The implications are numerous and important. First, neither Darwin is dispensable to his yet-imperfect theory. A Darwin who saw only categories would have been unable to see the *mutability* of species; a Darwin who saw only continua would have been unable to see the mutability of *species*. His abstract, however imperfect and *because of that imperfection*, needs both Darwins, even if their Janus-like conflation risks amalgamating incompatibilities, insisting at once on continuity and discontinuity.[34] Indeed, Darwin is always cutting the lines that he is otherwise always drawing: "In our diagram the line of succession is broken at regular intervals by small numbered letters marking the successive forms which have become sufficiently distinct to be recorded as varieties. But these breaks are imaginary, and might have been inserted anywhere..." (Chapter IV, p. 119). That required breaks are imaginary troubles Darwin no more than that they can be inserted anywhere. Indeed, this same gesture which he uses to articulate his theory – of a continuity reinterpreted as discontinuous (giving us distinct varieties and well-defined species), or a discontinuity reinterpreted as continuous (giving us a series "which impresses the mind with the idea of an actual passage")[35] – is the gesture which the theory itself articulates as unfurling in space and time: for example, in the cycles he describes of variety-engendering isolation (discontinuity), as of mountains separated by ice sheets, and stability-enforcing reconnection (continuity), when the ice sheets retreat (e.g. Chapter XI, pp. 365ff, especially p. 368-9; see also p. 353: "...geographical and climatal changes... must have

---

[34] One cannot help but think of Richard Dedekind's attempt to prove the continuity of the real numbers specifically by way of "cuts" (*Schnitts*) – an irrational number separating the reals into two sets, with all the members of one set being greater (or less) than those of the other set.

[35] A further example being insect castes which, like one perspective of species, are "perfectly well defined" (Chapter VII, p. 238), though Darwin will account for their evolution by revealing (from a different perspective) their "perfect gradations" (p. 239), such that "though the workers can be grouped into castes of different sizes, yet they graduate insensibly into each other..." (p. 241).

interrupted or rendered discontinuous the formerly continuous range of many species...").

Thus, rather than dwell on the reality of categories defined by breaks which he acknowledges are imaginary, Darwin, in two gestures both reminiscent of calculus, instead rationalizes them by understanding their series in terms of time and motion. First, as already suggested by his invocation of an "idea of an actual passage," he reads a muddled synchronic snapshot as representing a diachronic movement; and similarly, what on a diagram is an imagined break which could be placed anywhere is, in a no less imagined historical reconstruction, the advance and retreat of ice sheets. Thus the discontinuities that alone in his theory legitimize the categories by which we think are understood as the outcomes of continuous processes. Second, he insists on a disproportion between the timescale of change and our own possible experience. Categories, like the slowly-changing language in a volume of history imperfectly kept, are legible precisely because slowly-changing; on the timescale of our writing, as it were, they have the useful if ultimately false appearance of permanence. It is for this reason that Darwin not only repeatedly emphasizes gradualism but also unequivocally rejects catastrophism, as if the two perspectives necessarily were exclusive of each other. For Darwin, they are; in catastrophism, here understood broadly as the prospect of sudden and dramatic change, even the mere grafting of one plant onto another, lurks the threat of utter incomprehensibility, undoing in a single leap both gestures of rationalization, both assertions of the rational miscibility of continuity and discontinuity. More than once he confesses that the demonstration of the sudden emergence of, say, an organ or a species, without intermediate steps of gradual change, would be fatal to his theory (e.g., Chapter VI, p. 189; Chapter IX, p. 302; Chapter X p. 316). For, if species are to be well-defined for Darwin despite the term having an undiscovered and undiscoverable essence – or even, perhaps, despite having no essence – it is nonetheless on the condition of a history understandable as continuous in principle if not reconstructible as such in fact. Similarly, if calculus is to be comprehensible, whether in its geometric articulation by Newton or its numerical one enabled by Dedekind, it is on the condition that all changes can be approached continuously.

After all, whether in mathematical physics or in biology, *natura non facit saltum*.[36] Darwin repeats this Linnaean dictum not once but six times in The Origin. The gradual transition of a bear to a whale is thinkable (Chapter VI, p. 184), while the sudden transition of a bear to a kangaroo is "preposterous; for where there has been close descent in common, there will certainly be close resemblance or affinity" (Chapter XIII, p. 425).[37] Even when Darwin contradicts the dictum, it is only to

---

[36] "Nature does not make a leap."

[37] Yet, Darwin will qualify this claim, as he must, both by acknowledging beings that, though closely related, look extremely different (e.g. "three Orchidean forms... which had previously been ranked as three distinct genera" but were re-classified as a single species [Chapter XIII, p. 424] or the breeds of pigeons which "differ so extraordinarily... that they would, I cannot doubt, be ranked in distinct genera, had they been natural productions" [*Ibid.*, p. 445]) and, relatedly, especially with regard to the last example, by acknowledging that the conditionality of the reproductive system at times impairs its proper function: "... [variability] is due to the reproductive system being eminently sensitive to any change in the conditions of life,

affirm it at a different level: "On the theory of natural selection we can clearly understand the full meaning of that old canon in natural history, 'Natura non facit saltum.' This canon, if we look only to the present inhabitants of the world, is not strictly correct, but if we include all those of past times, it must by my theory be strictly true" (Chapter VI, p. 206). On that assumption, we are left only with two seemingly minor problems, each stemming from the second (and "more important") metaphorical sense of the term 'struggle for existence' and from the implicit confession therein, by the very designation 'metaphor,' that offspring are not perfectly like their parents. Indeed their resemblance, like everything else, comes in degrees (e.g., Chapter XIV, p. 475, on the complex laws governing the "degrees and kinds of resemblance to [offspring's] parents"). The two problems are these: how are monsters to be accounted for, and how close does a "close resemblance" have to be? Or, in other words, for either case, when is a saltational threat to be suspected?

On the one hand, Darwin will endeavor to domesticate saltation. This is what he does with monsters. Were a bear to

---

being thus often rendered either impotent or at least incapable of its proper function of producing offspring identical with the parent-form" (Chapter VIII, p. 273; see also Chapter XIV, p. 466). Such deviations Darwin will always understand as evidence of an unnatural state (as he does with the pigeons above) – even if he also has to rigorously admit that resemblance to one's parents, like everything else, comes in degrees and kinds (e.g., Chapter XIV, p. 475). Were it otherwise, there would be no mutability of species, no natural selection and no theory of evolution – outcomes which should challenge Darwin's understanding of the "proper function" of the reproductive system as also his distinction between natural and unnatural states.

produce an offspring like a kangaroo, that offspring would be understood as monstrous, i.e., as an assemblage of monstrosities, a monstrosity being defined by Darwin as "a considerable deviation of structure in one part, either injurious or not useful to the species, and not generally propagated" (Chapter II, p. 44). A naturalist expecting a bear cub would likely see a pouch as injurious or at least useless – forgetting that in the evolutionary play of context neither use nor injury is a fixed term – and might accordingly ascribe the whole matter to "the unnatural treatment of the embryo," an interpretation, Darwin notes, supported by the experiments of St. Hilaire (Chapter I, p. 8). By considering the 'kangaroo' monstrous, any threat this seeming saltation poses can be dismissed. Monstrosities occur in nature, as Darwin, like everyone else, knows, but they are exceptions, a form of nature gone awry, the result of an "unnatural treatment." Yet, Darwin *also* acknowledges – and this is more surprising – that sometimes, as in the case of hybrids and mongrels, resemblance to one's parent may be "chiefly confined to characters almost monstrous in their nature" (Chapter VIII, p. 275). Monstrosity, which would normally be a dramatic break in resemblance between parent and offspring, can instead be the conveyor, and even the chief conveyor, of that resemblance, albeit again in the supposedly unnatural case of hybrids or mongrels[38] and despite "not generally [being] propagated." Such an acknowledgement may make monstrosities less threatening, but in the process it has

[38] Or, as we shall see in a moment, the domesticated breeds whose methodical selection provided the analog according to which natural selection was hypothesized.

unsettled the entire theory of "close resemblance or affinity" paired to close descent – since here resemblance is "chiefly" only in the monstrous or almost monstrous characters. What of all the other characters?

But Darwin doesn't stop there. For he has already admitted, in a version of the logic he has so often deployed, that monstrosity exists on a continuum that includes less extreme forms: "monstrosities cannot be separated by any clear line of distinction from mere variations," he writes at the very start of the book (Chapter I, p. 8). He re-emphasizes the point at the beginning of Chapter II (p. 44): "We have also what are called monstrosities; but they graduate into varieties."[39] But if that's the case, our problem has been solved at the mere expense of re-creating it. The threat that saltation posed to his theory has been tamed by incorporating it into the continuum that formerly it broke. Very well; but then we no longer have an *a priori* reason to reject as preposterous the kangaroo born of the bear, which, after all, differs only in degree from a bear rendered by natural selection more and more aquatic until "a

---

[39] Likewise, Darwin admits that "domestic races of the same species, also, often have a somewhat monstrous character" (Chapter I, pp. 15 – 6), even while he later acknowledges that, excepting perfectly fertile mongrels and these monsters, "domestic races of the same species differ from each other in the same manner as, only in most cases in a lesser degree than, do closely allied species of the same genus in the state of nature" and that many competent judges disagree about whether to rank such races as varieties or species (p. 16). If he is to escape the bind of a continuum between monstrosity and, ultimately, species, it will be on the grounds of an absolute distinction between man and nature, between what is domestic and what is natural – but he problematizes this distinction no less than any other, most obviously in the analogy he makes between methodical and natural selection.

creature was produced as *monstrous* as a whale" (Chapter VI, p. 184, italics added).

On the other hand, Darwin would simply avoid saltation by emphasizing a preponderance of really small steps, if not visible now, at least formerly existent. His calculus-like hope is that what from one vantage looks like a staircase from another resembles a ramp. Thus, bearing in mind that the breaks between successive forms are imaginary, we nonetheless confront what is, for Darwin, literally a flight of steps: "Hence I look at individual differences... as being the first step towards such slight varieties as are barely thought worth recording in works on natural history. And I look at varieties which are in any degree more distinct and permanent, as steps leading to more strongly marked and more permanent varieties; and at these latter, as leading to sub-species and to species. The passage from one stage of difference to another and higher stage... I attribute... to the action of natural selection..." (Chapter II, pp. 51-2). But how big can a step get before it requires a leap? In a different context, Darwin confronts this problem when he rejects geographic distance as a criterion by which to distinguish doubtful forms as distinct species. For, he notes with sublime simplicity, "what distance, it has been well asked, will suffice?" That between America and Europe? What about the distance "between the Continent and the Azores, or Madeira, or the Canaries, or Ireland" (Chapter II, p. 49)? Distance is always relative; but so too are the sizes of steps. Darwin never confronts the symmetry of transformation, the niggling fact that a staircase transformed into a ramp can also be seen as a ramp transformed into a staircase. And so he also fails to see that the continuity by which he would arrive at thought-enabling discontinuities is itself riddled – at every step – with discontinuities.

Put otherwise: Darwin wants simultaneously to privilege being – implicitly understood as presence, though always as it were accidentally hindered from its full realization – and becoming. While he recognizes the explosive potential of this combination, he seems to hope that a titrated admixture in small enough doses will indefinitely defer that eventuality. As if any becoming of being, however gradual, weren't already catastrophic – much more a becoming without beginning or end.[40] His predicament is akin to Newton's in the latter's

[40] Here we might consider the Aristotelian privilege given the adult form, which thus sanctions a becoming into being. Darwin frequently accepts this view, albeit in terms explicitly dramatic: e.g., "But it is by no means obvious, on the ordinary view, why the structure of the embryo should be more important for this purpose [of classifying] than that of the adult, which alone plays its full part in the economy of nature" (Chapter XIII, p. 418) and again (Chapter XIV, p. 480) "But disuse and selection will generally act on each creature, when it has come to maturity and has to play its full part in the struggle for existence..." Yet, as so often with Darwin, he can't help but qualify his agreement: "The case, however, is different when an animal during any part of its embryonic career is active, and has to provide for itself. The period of activity may come on earlier or later in life; but whenever it comes on, the adaptation of the larva to its conditions of life is just as perfect and as beautiful as in the adult animal" (Chapter XIII, p. 440). Of course, that larval 'perfection' may, through the "imperfectly understood" laws of correlation of growth, impact the adult form, "modifying important structures, independently of utility and, therefore, of natural selection" (Chapter V, p. 144) – and hence, on Darwin's logic, may jar the adult form away from perfection. Adaptive characteristics at particular life stages may have correlates at other life stages that are nonadaptive or even, potentially, maladaptive. Doesn't this threaten to destabilize the imagined perfection embodied by the mature form? Or at least to call into question what

development of calculus, not least because both are trying to give an account of motion, of continuous becoming, in terms of discrete, hence discontinuous, beings. Each would, as it were, contain and hence localize the incendiary danger by understanding its unfolding as simply temporal and/or spatial – time and space being, as supposedly inert tableau, quarantined from the dynamic they would encompass. Becoming, like the necessary imperfection of Darwin's abstract, would thus always arrive as an accident and hence would not challenge the priority of being both initially and finally. Hence the unexamined postulates of both an origin and an end – postulates required not by Darwin's argument but by his metaphysics. If nothing else, a discrete beginning

---

perfection could possibly mean? Further, is it really clear what counts as activity in the career of an embryo, or would careful thought on such activity expose all embryos as, in one form or another, active? Finally, while Darwin writes in Chapter XIII (p. 441) that "no one probably will dispute that the butterfly is higher than the caterpillar," does that really mean we shouldn't? Especially when even Darwin has to admit that "it is hardly possible to define clearly what is meant by the organization being higher or lower" (Chapter XIII, p. 441)? Perhaps we should instead examine why it is that we privilege adult over juvenile forms, or even heights over depths. At issue ultimately is the *impossibility* of being adapted to all possible contexts, even if context always bounds meaning. Privileging an adult form would vainly endeavor to privilege a particular and proper context. But there is no limit to contexts. There is also (correspondingly) no stark divide between juvenile and adult forms; hence the mite *Acarophenax tribolii*, mentioned in Chapter 1, footnote 4, a juvenile (male) form of which might be understood as an adult (female) organ. The physiological division of labor to which Darwin repeatedly turns also problematizes the distinction between organism and organ, since organisms in a colony of insects (pp. 241 – 2) or in an ecosystem (pp. 115-6) become specialized in this way no less than organs (p. 93 and pp. 115-6).

and end mirror the discontinuities necessary for thought-enabling distinctions, for *definition*, by excising from a processual blur individual things. Thereby would Darwin safeguard being – it is, after all, what Nature tends in a mode which Darwin insists is ultimately preservative rather than destructive – even as he describes its indefinite transformations and repeated annihilations. But that indefiniteness, that deferral of the end – in all senses of that word – like the parallel deferral of the beginning, *is* the problem, one less readily instrumentalized in biology than in calculus: for infinity might look like a mere concept, but life looks like what we are.

This unholy alliance, however, between being and becoming jeopardizes concepts no less than what we are. To advocate abandoning the search for the undiscovered and undiscoverable essence of the term 'species' is to abandon a philosophic project since Plato. Consider, for example, this discussion of the essence of bees in *Meno* (72b-c, my translation); it will serve, Socrates promises, as the ostensible template for determining the essence of any term and, in particular, of the term aretē (ἀρετή, virtue, excellence).

Socrates: But, Meno, according to this image of swarms, if I were to ask you what the essence (οὐσία) of a bee is, and you said that bees are many and various, what would you answer me if I went on to ask, 'By this are you saying that they are many and various and different from each other insofar as they are bees? Or, in being bees, do they not differ from each other at all, whether in terms of beauty or size or some other such thing?' Tell me, what would you answer if you were thus questioned?"

Meno: "I would say this, that they do not differ at all the one from the other insofar as they are bees."

Socrates: "If, after that, I were to say, 'Tell me, then, in what, Meno, do they not differ in any way but are all the same, what would you say this is?' Would you perhaps have anything to tell me?"

Meno: Egōge (ἐγώγε).

Three points merit highlighting. First, metaphor is at the heart of Socrates' question. He invites Meno to unite different things (individuals we call bees) according to their shared commonality – here, explicitly their shared essence, whatever it is which makes them, though different, all bees. But is metaphor an implicit recognition of a shared essence, or is it instead the condition of possibility on which rests the postulate, declaration or disclosure of essences? Second, despite Darwin's own dependence on metaphor, observe that his theory of evolution is *also* predicated precisely on bees differing *even insofar as they are bees*. So too every other organism. The mutability of species both admits and relies on that difference. Finally, note that Meno's enigmatic answer, ἐγώγε, which here means something like, "for my part" and is taken both by Socrates and, universally, by translators as an affirmation (perhaps best translated "Aye"), is nothing more than this supposed affirmation. Meno and Socrates do not follow up this assertion with its elucidation; what remains shared between the two men is their agreement that the essence of bees *could* be articulated – *but they do not bother to do it*. Likewise, despite by the end of the dialogue failing to agree on the essence of the term ἀρετή, neither man as a result embraces the Darwinian conclusion that the search is vain.[41] To do so, as far as Plato's Socrates is concerned, would destabilize language, thought and inquiry. Which is precisely the charge laid against certain formulations of the theory of evolution by its later critics, none better than G. K. Chesterton (Orthodoxy, pp. 30-1):

"Evolution is a good example of that modern intelligence which, if it can destroy anything, destroys itself. Evolution is either an innocent scientific description of how certain earthly things came about; or, if it is anything more than this, it is an attack on thought itself. If evolution destroys anything, it does not destroy religion but rationalism. If evolution simply means that a positive thing called an ape turned very slowly into a positive thing called a man, then it is stingless for the most orthodox; for a personal God might just as well do things slowly as quickly, especially if, like the Christian God, he were outside of time. But if it means anything more, it means that there is no such thing as an ape to change, no such thing as a man for him to change into. It means that there is no such thing as a thing. At best, there is only one thing, and that is the flux of everything and anything. This is an attack not upon the faith, but upon the mind; you cannot think if there are no things to think about."

[41] But neither entirely does Darwin, as my previous comments should make clear.

Certainly, the flux of everything and anything destabilizes the categories by which we think. This is the threat that Darwin is ever confronting and resolving through the imposition of what he rigorously acknowledges, at least at times, are imaginary breaks. But where Darwin's rigor exceeds Chesterton's, albeit perhaps despite Darwin and without his awareness, is in his repeated recognition that the flux of everything and anything is also *the condition of possibility* of categories. Not one but two threats menace the mind: you cannot think if there are no things to think about, but neither can you think if there are *only* things to think about. The cacophony of absolute particularity, what Aristotle called primary substance in the Categories, is babble or, in another sense, Babel. Communication and thought are impossible in a world exclusively of proper nouns. It was this admission which drove Aristotle to the invention of secondary substance, without which he literally had nothing to say. Only when a thing can join a series of like things, only when it can be mobilized rhetorically by simile, metaphor or metonymy, can that thing, in some enormously qualified sense, be thought, albeit at the risk of impressing the mind with, e.g., the idea of an actual passage – the idea that an unbridgeable chasm has indeed been crossed.

Thus, to emulsify being requires a tincture of becoming. But such an emulsion, the achievement of which is seemingly the thinkability of things, also destabilizes its own ingredients. No longer can being and becoming be absolutely distinguished. Something must therefore be invoked to limit and regulate what otherwise risks being a chain reaction. For Chesterton, it is the Christian God. For Plato's characters, it is at least the concession of an eventually articulable essence. For Darwin, it is an origin, an end, a place or a Creator. Regardless of what name is given to it, it has, inevitably, the characteristic of a presence uncontaminated with becoming; it alone is fully itself; it alone is fully present; for it alone is being fully presence. This would be the transcendental signified where signification both begins (origin, archē, ἀρχη) and ends (perfection, telos, τελος). But this is a desire, not an achievement. In particular, in the case of Darwin, an origin, however often invoked, is ever-receding; an end, though regularly promised, remains forever uncertain; and his Creator, making (in the first edition) but the occasional appearance throughout the abstract,[42] is scandalously absent in the book's concluding

[42] Darwin mentions the Creator seven times in the first edition. Four of those references are in the context of the arguments of others, e.g. "he who believes in separate and innumerable acts of creation" (Chapter VI, p. 185; see also Chapter XIII, pp. 413 [2 examples] and 435). The seventh (in Chapter XIV, p. 488) is sufficiently instructive that I shall discuss it in the main text. The other two are in Chapter VI, pp. 188-9, specifically in the context of a comparison – famous at least since Paley's Natural Theology (1809) – between telescopes and eyes. Here Darwin proceeds with considerable nuance. The inference "that the eye has been formed by a somewhat analogous process" [as that which "perfected" the telescope] strikes him as "presumptuous": "Have we any right to assume that the Creator works by intellectual powers like those of man?" Instead, he describes a process in which "transparent tissue, with a nerve sensitive to light beneath" is continually modified; and he urges that we suppose "a power always intently watching each slight accidental alteration... and carefully selecting each alteration which, under varied circumstances, may in any way, or in any degree, tend to produce a distincter image." Scale this up to millions of instruments, with preservation of the better and destruction of the worse; and let "this process go on for millions on millions of years." "[M]ay we not believe," he concludes, "that a living optical instrument might thus be

sentence (as discussed earlier, pp. 111). As much as Darwin relies on a transcendental signified, he also unsettles it. Much of the historical appropriation of Darwin has proceeded by endeavoring to quell that unsettlement, frequently according to terms that Darwin's text had foreseen, dallied with – and rejected. For example, the deployment of reproductive compatibility and isolation in order to define species would become the basis of Mayr's "biological species concept," despite the objections, discussed in Chapter 1 (pp. 58 – 9), that Darwin all too rigorously raised. Likewise, the discovery of DNA and its structure led to an extraordinarily reductionist account of life, still prevalent today, that all but ignores the role of mutual relations in shaping living things. Instead, all that mattered was the second metaphorical sense of the term 'struggle for existence,' success in leaving progeny. But the ascendancy of DNA led to the discovery of horizontal gene transfer, the commerce in genes among neighbors irrespective of their parentage (with the consequence, among others, that the concept of "parent" loses its assumed self-evidence). As a result, the first metaphorical sense re-asserted itself as co-equal

---

formed as superior to one of glass, as the works of the Creator are to those of man?" Note the subtle coup that Darwin has achieved: for we still have a comparison ostensibly between eye and telescope, works of Creator and man respectively; but if these are works of the Creator they are also works of variation (which "will cause the slight alterations"), generation (which "will multiply them almost infinitely") and natural selection (which "will pick out with unerring skill each improvement"). Would these be the instruments of the Creator or His attributes? Or, in taking the place of the Creator, would they testify to His theoretical redundancy, the absence of a need to invoke Him? Can we be certain that the Creator is any less absent in this otherwise direct invocation of Him, than He is in the last sentence of Darwin's abstract?

with and ultimately inseparable from the second, and thereby ecology necessarily transgressed into biology. The deeper problem, the commitment to the metaphysics of presence which these mistakes cannot help but simultaneously to affirm and problematize, assuredly was not seen by Darwin's heirs. Just as is the case with Darwin's own text. Darwin is like his own intellectual progeny in trying to tame the wild implications of his theory;[43] and those progeny are like him in forever unsettling the very thing they believe themselves to have at last firmly established. Their insight always begins at its own limit; and the borders between what is seen and what is not prove to be in play, whether because of man's inventiveness (with the telescope) or nature's (with the eye). No less imaginary than the breaks on Darwin's diagram or the boundary between his two metaphorical uses of the term 'struggle for existence' is the border between seen and unseen.

---

[43] Thus the human being is responsible both for tameness and wildness, whether theoretically (as here) or empirically, as with animals ("Nothing is more easy than to tame an animal..." Darwin writes in Chapter I, p.8, and again p. 17, "...but how could a savage possibly know, when he first tamed an animal..." – thereby accepting taming as an act of man; but wildness is also due to human action: "But fear of man is slowly acquired... and we may see an instance of this, even in England, in the greater wildness of our large birds than our small birds; for the large birds have been most persecuted by man" [Chapter VII, p. 212]). Human responsibility, however, is far from reducible to a pure and active agency. Rather human agency is revealed as neither simply active, nor simply passive (as already suggested by the implied limitations above of a savage's knowledge) – and the human attempts to tame the wild implications of a theory don't fail to contribute to its wildness. This admixture of agencies, which is a consequence or at least a correlate of the emulsion of becoming and being, in short of finitude, is discussed in greater detail in Chapter Three.

Though legislated, declared or envisioned, it cannot be found.[44]

IV. Rudimentary conclusions and the organs of writing

"Rudimentary organs may be compared with the letters in a word, still retained in the spelling, but become useless in the pronunciation, but which serve as a clue in seeking for its derivation." (Chapter XIII, p. 455)

"Rudimentary organs will speak infallibly with respect to the nature of long-lost structures." (Chapter XIV, p. 486)

Rudimentary organs represent one of the many seams where Darwin's text is stitched together. He understands them as strong support for his theory over and against "the ordinary doctrine of creation," for how can the latter explain "the existence of organs in a rudimentary, imperfect, and useless condition, or quite aborted" (Chapter XIII pp. 455-6)? Further, they are of "high classificatory value," being – but should this not strike us as strange? – among those "most permanent characters" by which the lines of descent are reliably traced (Chapter XIV, p. 479). Their meaning, Darwin insists, "is often quite unmistakable" (Chapter XIII, p. 451). Yet, that meaning is also always in play: the "variability [of rudimentary parts]

44 Hence the importance, already noted in Footnotes 22 and 29, of our qualified vision when regarding what may appear to be absolute perfection: "Beyond this stage of perfection in architecture, natural selection could not lead; for the comb of the hive-bee, *as far as we can see*, is absolutely perfect in economizing wax" (Chapter VII, p. 235).

seems to be owing to their uselessness, and therefore to natural selection having no power to check deviations in their structure. Thus rudimentary parts are left to the free play of the various laws of growth, to the effects of long-continued disuse, and to the tendency to reversion" (Chapter V, pp. 149-50). However useless, these organs, precisely because subject to such free play, also become the nursery of innovation: "[A]n organ may become rudimentary for its proper purpose, and be used for a distinct object: in certain fish the swim bladder seems to be rudimentary for its proper function of giving buoyancy, but has become converted into a nascent breathing organ or lung" (Chapter XIII, p. 452).[45] Yet, the propriety of purpose, far from being put into play by the play of

[45] Darwin gives the same example in Chapter VI (pp. 190-1), albeit omitting reference to rudiments – though their possibility remains legible in the earlier discussion (also on p. 190) of the similar functions of gills (or branchiae) and swim bladders, a redundancy enabling perfection of one organ and either modification to another purpose or obliteration of the other. Regarding the relationship of swim-bladder to lung, Darwin notes the plurality of its purposes: "originally constructed for one purpose, namely flotation," it was "converted into [an organ] for a wholly different purpose, namely respiration." (He also notes the co-implication of the swimming bladder and a fish's auditory apparatus, though he is unsure which is complementary to which – a confusion which should give him pause.) In this light, "there seems to me to be no great difficulty in believing that natural selection has actually converted a swimbladder into a lung, an organ used exclusively for respiration." Among the difficulties he doesn't see are those that now should plague his use of the adverbs "originally" (was a swim-bladder "originally constructed for one purpose" or are its origins like the lung's?) and "exclusively" (do lungs have the exclusive purpose so elusive for the swim bladder?), not to mention his general faith in purpose and perfection and his assumption, however qualified, of a context-independent meaning of either organs or organisms.

rudimentary parts, is paradoxically preserved by it: "…the same reasoning power which tells us plainly that most parts and organs are exquisitely adapted for certain purposes, tells us with equal plainness that these rudimentary or atrophied organs, are imperfect and useless" (Chapter XIII, p. 453). Likewise, although "[n]othing can be plainer than that wings are formed for flight" (Chapter XIII, p. 451) – even if ostriches use them "as sails" and penguins "as fins in the water and front legs on the land" (Chapter VI, p. 182) – the membranes found in flightless beetles must nonetheless be "rudiments [that] represent wings" (Chapter XIII, p. 451). They represent what they are and are not, and thereby place Darwin at odds with himself: even as he *describes* an innovative utility of rudimentary parts, he fails to *perceive* them as anything other than useless; and in lock-step he fails to perceive the threat to proper purposes that his account cannot fail to describe. He can go so far as to acknowledge without trembling that "the rudiment [of a pistil] and a perfect pistil are essentially alike in nature" (Chapter XIII, p. 451).

On the one hand, he wishes to distinguish, as context-independent characteristics, the rudimentary from the perfected, the useless from the purposive. On the other, he recognizes, in a certain sense despite himself, that they are "essentially alike." That likeness too would be independent of context, even if no one more than Darwin has stressed the role of context in determining the always-provisional meaning(s) of organic beings, including, not only of their 'functional' parts, but of their rudimentary ones. But what could this essential likeness be? It is not a likeness of function, clearly; nor of form, since it is no small inference to see represented in the "mere rudiments of membrane" a wing – a conclusion possible only through comparison to other organisms (Chapter XIII, p. 451). Indeed, what is it to be essentially *alike* rather than essentially *the same*? Perhaps that is the clue, a sameness in only ever approaching likeness: that is, what they share 'essentially' is the protean capacity of each to transform into something like the other, the perfect into the rudimentary no less than the rudimentary into the perfect. But this is quite simply to challenge the priority and even possibility of essences. Of course, one could object that Darwin is using the word "essentially" only in a vulgar sense, but – setting aside the unavoidable irony of such an objection – that vulgar sense, even if unsophisticatedly, still clings to the possibility which it declares, the possibility which Socrates and Meno agree to accept with regard to bees and then immediately defer. It is the promise that the play of what is never quite assured as meaning will end – somewhere: in the essence or being (οὐσία) of a bee or of a wing or of a pistil; in absolute perfection; in a triumphant teleology ultimately indistinguishable from a cataclysmic eschatology, where history, long imperfectly kept in a volume of changing forms, simply ends and life achieves the absolute stagnation of full, unstaggerable and irreplaceable presence.

In parallel, Darwin confuses speech and writing. The latter in particular has frequently furnished him with figures for life and by which to understand life – whether (following Lyell) in terms of that last volume of history (Chapter IX, pp. 310-1 and again in Chapter X, p. 333), of a slowly changing dialect (Chapter I, p. 40; and note that Lyell's metaphor reveals writing as the mode of revelation of such changes), of Bacon's "book of God's works" quoted as an epigraph to The Origin (p. ii), of a tree-like diagram "almost impossible to do... without" (Chapter XIII, p. 431), of an embryo as a palimpsest-like "picture, preserved by nature, of the ancient and less modified condition of each animal" (Chapter X, p. 338), or even simply of a "vertical line of varying thickness" (Chapter X, pp. 316-7 and again 318). Yet Darwin has never confronted *why* he has so often been driven to writing, as indispensable tool and figure, in order to think about life. He has never wondered about the *absence* always marked by writing. Instead, he slips into a dream of that moment when writing will at last be displaced by a speech become faultless, unfailing, perfect. It is a dream in which the play of representation ends and plain signification, notably in speech, at last prevails:

"Naturalists frequently speak of the skull as formed of metamorphosed vertebrae: the jaws of crabs as metamorphosed legs; the stamens and pistils of flowers as metamorphosed leaves... Naturalists, however, use such language only in a metaphorical sense: they are far from meaning that during a long course of descent, primordial organs of any kind – vertebrae in one case and legs in the other – have actually been modified into skulls or jaws. Yet so strong is the appearance of a modification of this nature having occurred, that naturalists can hardly avoid employing language having this plain signification. On my view these terms may be used literally..." (Chapter XIII, pp. 438-9)

Darwin fails to see that such plain signification is achieved at terrible cost: legs which become jaws and leaves which become pistils can only throw into doubt the plain signification of jaws, legs, leaves and pistils. "Everything is leaf!" Goethe ecstatically proclaimed as summary of his theory of plant metamorphosis in a 1787 letter to Charlotte von Stein, leaving unstated its corollary, that nothing is, not even a leaf. Leaving unstated what could only be written: that "leaf" now had to be understood as perpetually signifying, never merely signified. But that is what writing is: a mark always of absence; a 'system' of signs whose referentiality fails to be exhausted in a final signified and hence fails to be systematized; a signified always revealed as but the delicate chrysalis of a signifier; a signifier, then, of signifiers.

Darwin, as so often, does and does not see this. Or rather, synaesthetically, he would hear what must instead be seen. For what is legible in his text, writable and readable, is nonetheless not sayable. Thus, only a few pages from the end of his imperfect abstract, he promises a coming day when "rudimentary organs will speak infallibly" (Chapter XIV, p. 486); yet these same organs are those which he compares to "the letters in a word, still retained in the spelling, but become useless in the pronunciation" (Chapter XIII, p. 455). Letters of a word, in brief, which can be written and read but not spoken.[46] For how do such letters, the 'b' in 'doubt' for example, *speak* infallibly? The letters themselves, like the

[46] Have we not somehow again stumbled upon a striking resemblance to that structure of deferral and difference that destabilizes the organizing oppositions of our thought, including inside/outside, metaphor/metonymy, signified/signifier, and that Derrida nicknamed *différance*, distinguishable from its homophonous *différence* only by the legible substitution of an 'a' for an 'e'?

rudimentary organs, cannot speak; and the one speaking for them is forced into a decision: shall the 'b' in 'doubt' be articulated, though that is to say a different word than 'doubt'? Or shall that 'b' be left unarticulated, thus introducing the possibility, past, present and future, of fallibility, of a 'mispronunciation' become accepted? Yet Darwin is not startled here by a sense of impending contradiction, as he is not by the parallel problems discussed already in his deployments of place, perfection, essence, origin, etc. Darwin's great insight in The Origin, from which all else follows, is to read life as writing; and, though this is an insightful blindness, to stop. Writing, always a mark of absence, would nonetheless also be, for Darwin, the mark of something eventually present. Recourse to writing, literally or figuratively, is accordingly an accident, much like his present inability to provide references for his consequently imperfect abstract. Understood simply as a descriptive tool, writing merits no further thought beyond what utility it has achieved, servant ultimately to a meaning whose eventual full presence Darwin can neither doubt nor fail to defer. Hence his recourse a second time to "plain signification," albeit again as a promise deferred to the future, the fruit (but this fruit too is leaf) of his own theory:

"The terms used by naturalists of affinity, relationship, community of type, paternity, morphology, adaptive characters, rudimentary and aborted organs, &c., will cease to be metaphorical, and will have a plain signification" (Chapter XIV, p. 485).[47]

---

[47] The only other uses of "plain signification" in The Origin are both enlightening: In Chapter II (pp. 55-6), the "plain signification" of an

Yet he cannot help himself, and the very next sentence leaps into an almost epic simile, beginning: "When we no longer look at an organic being as a savage looks at a ship…" For there is nothing plain about the signification of these words, any more than there is of his use of the term 'struggle for existence.' Indeed the two metaphorical poles of the latter are often what put "plain signification" and the "terms used by naturalists" into play. Will a "relationship" mean that of descent (genealogical relations) or of dependency (mutual relations)? Which of these will morphological likeness represent? Are adaptive characters truly to be distinguished from the rudimentary, or will the latter, however uselessly inherited, become contextually useful? And when, Stephen Dedalus might ask, has paternity ever been a matter of plain signification? The organic being who confronts us and who we are is protean, one moment like a ship (and more than that: like a ship seen by a savage), at another like a wreck (Chapter V, p. 139),[48] at yet another, like "mariners shipwrecked near a

---

assemblage of facts comes in view solely through the acknowledgment "that species are only strongly marked and permanent varieties" (of course "permanent," it can only emerge, here means anything but permanent) while in Chapter XIII (p. 435) the "plain signification of the homologous construction of the limbs throughout [a] whole class" reveals the link uniting, e.g., fin, webbed foot and wing. Both examples are but Goethian leaves in disguise.

[48] Speaking partially in reference to the "blind Proteus" found in caves – i.e., the amphibian *Proteus anguinus*, also known as the olm – Darwin writes, "I am only surprised that more wrecks of ancient life have not been preserved, owing to the less severe competition to which the inhabitants of these dark abodes will probably have been exposed."

coast" (Chapter V, p. 136) – all and none.[49] "Proteus" indeed is an apt name, like "leaf"; and our difficulty, mirroring that of Menelaus in the Odyssey, is precisely in wrestling a being who has no final and no original form. As we also do not: for it is no minor detail that the man who entered Troy disguised as a horse ambushes Proteus in the likeness of a seal, nor that the Socrates ostensibly seeking the essence of a bee is in form most like a sting-ray at one time and a gad-fly at another.[50]

To claim that Darwin's insights can comprehensively be summarized as an understanding of life as writing, an understanding the endlessly unfurling implications of which Darwin himself cannot sustain, is simply to read him. In the final few pages of his book, the Creator at last makes an appearance. As He was at the beginning of the book, in the epigraph that Darwin cites from Bacon, so too is He at the end – an author responsible for "the laws impressed on matter" (Chapter XIV, p. 488), whether that matter is living, nonliving or dead.[51] From alpha to omega, the Creator would impress, as with a chisel or a stylus or merely with a finger, laws into stone

---

49 I cannot help but think of Wordsworth in *The Prelude* (1850), V 120-5: "*– Lance in rest, / He rode, I keeping pace with him; and now / He, to my fancy, had become the knight / Whose tale Cervantes tells; yet not the knight, / But was an Arab of the desert too; / Of these was neither, and was both at once.*"

50 See *Meno* 80a and *Apology* 30e, respectively.

51 Notably our propensity to *read* a series of gradations as representing an actual passage is something *impressed* on our minds (Chapter II, p. 51: "...a series impresses the mind with the idea of an actual passage"). Perhaps under the influence of Hume, Darwin uses this image of the mind as a writing surface, a surface to be stamped, regularly (e.g. Chapter IX, p. 283: "With the mind thus impressed..." or p. 285: "The consideration of these facts impresses my mind almost in the same manner as does the vain endeavor to grapple with the idea of eternity.")

or clay, tablet or organic being; and "by the establishment of general laws" thus is obviated the necessity of invoking "insulated interpositions of Divine power," in the words Darwin quotes from Whewell as his other book-opening epigraph. Readied is the way for the absence of the Divine; what He leaves in his stead is a text. A text that calls into questions ends and origins, hence even Creators, in a play of signification that culminates in a signified only by fiat or by faith. Darwin's own text approaches being an allegory of this other text, a writing which cannot be traced, much less reduced, to the intentionality of an author: a writing which, in being read, reveals what an author could not say, much less breathe into a few forms or into one. And in parallel, those other authors – the forgetful ones with which this chapter began, the ones who would control evolution – imagine themselves as writers who *say* only what they mean – writers, in short, who know neither what writing "is" nor that they are writing.

What writing betrays is life: in the revelatory mode of the innermost secret exposed like a purloined letter in plain sight, the substance of which is insubstantial, the essence inessential; and in the mode of disloyal abandonment, whereby the principle of death and of différance is or becomes legible in the becoming of being.

Parabasis (παράβασις): A Little More than Kin and Less than Kind – A Dumb-show Interlude

Belike this show imports the argument of the play

Let me begin by calling attention to two implications of Darwin's theory of evolution. First, living beings are related to one another and possibly stem from a single origin.[1] Second, in part as a consequence of that relatedness, distinctions of kind, notably of species, become problematic. Indeed, as already discussed, Darwin explicitly renounced the search for the essence of the term 'species' as "undiscoverable" in On the Origin of Species. In the context of these implications, my question is simply, what does it mean for living beings to be kin even while they are less than kind? I shall sketch two closely-related theses in answer to this question. I shall try to show that both orbit the broader problem of purpose or end – what I will call telos (τελος) – and the metaphysics of which it is an expression.

I. Proteus

My first thesis is that evolution, as an idea as well as biological phenomenon, challenges the categories of our language and thought.[2] It suggests that they are at best

---

[1] "I believe that animals have descended from at most only four or five progenitors, and plants from an equal or lesser number. Analogy would lead me one step further, namely, to the belief that all animals and plants have descended from some one prototype." On the Origin of Species, Chapter XIV, p. 484. For a critique of these implications, see Chapter 2.

[2] To quote Woese and Goldenfeld (2009. p. 17), biology in the wake of evolution emerges as "a study, not in being, but in becoming."

provisional.[3] In fact, it places us in a position akin to Menelaus' when he wrestles Proteus, that old and possibly divine being who, in an effort to escape Menelaus' grasp, twists and turns into every beast that crawls the earth and into water and fire as well. Yet Menelaus holds on. To what? What are we to grasp, what are we to comprehend, in a protean world? After all, in contrast to Proteus' metamorphoses, it is not clear that evolution has an end. Menelaus and his men, if only they are tenacious enough, can outwait Proteus. The god, curiously, tires. He reverts to a final form, the kind that – at least for a while – he is. But what would a final form of life be? When does evolution stop and allow life to rest in a final, true or original form? What will that end, that τελος, have been, the one towards which we and it have ever been twisting and turning? What if it's not there? Can we honestly think *that* world?

This last question is routinely answered in our modern age with little more than an expressive shrug. It's a commonplace now to abandon τελος. Indeed, the teleological biologist is a heretic. But I suspect he's also extremely pious. I suspect that when modern biology sidesteps the problem of τελος by invoking, for example, the language of function –

---

[3] Adam's naming of the animals in Genesis is then the more mysterious. It seems that, through Adam's pronouncement, animals start with proper names; how do they become common? How do they stick to organisms that change both developmentally and generationally? It's interesting in this context to wonder why so many characters in Genesis change their names (Woman/Eve, Abram/Abraham, Sarai/Sarah, Jacob/Israel, not to mention the many attempted names of God) and whether Adam himself ever received a name. If he did, how, from whom, and is it a common or a proper noun? No less mysterious is how a common noun becomes proper.

eyes, we initiates say, have functions, not purposes – it has only impoverished, not abandoned, the teleological orientation that seemingly makes the world graspable. After all, it is at least partly through the extension of a mechanical metaphor that eyes are suddenly recognized as instruments with functions. The embrace of that metaphor necessarily entails an embrace of instrumentality, of understanding organs and even living beings as means.[4] But means toward what? Have we really escaped teleological thinking if we simply don't offer an explicit answer to that question? Indeed, the question is implied in the mechanical metaphor from the very beginning: after all, would anyone say that the function of your standard machine – your iPod, your elevator, your windmill – isn't purposeful?[5]

In a protean world, a teleological biology is confronted with the irritating way that seeming ends keep turning into something else; yet it holds on in the expectation of a final end, a true τελος, a god – or is it merely an old man? – in hand. A mechanistic biology does not avoid a similar cascade of meaning: evolution challenges the stability of function no less than of purpose. Just as a hammer can prop up a window as well as pound a nail, so too can an esophagus develop the function of a lung. In short, both the teleological and mechanistic biologies confront a world which they characterize not by the absence of meaning but by its seeming

---

[4] It is common to discuss, for example, the function of an individual (as the production of progeny, say), or the ecosystem function of a given species.

[5] This is of course different from claiming that it is only purposeful. For the secret rapprochement of mechanism and teleology, see Chapter 1 of Bergson's Creative Evolution.

superabundance. The mechanistic response to that problem is the very gesture of measurement: the enormously powerful, metonymic reduction of the indefinitely complex object to a privileged, quantifiable trait. As soon as that trait is confused for the object, as it inevitably is, the mechanistic biology has its τελος: a world seemingly overrun with meaning becomes instantly graspable at the mere expense of its own impoverishment, and the wonder of sight is replaced by its shadow, the function of an eye.[6] In this way, modern biology, in its invocation of function instead of purpose to explain evolutionary developments,[7] has not escaped the problems or even the implicit supposition of τελος, even if that τελος remains undisclosed. It has avoided the word and, largely by way of an unexamined rhetoric,[8] hollowed out its possible

---

[6] This now-graspable-but-impoverished world is nonetheless *still* pervaded with more (apparent) meaning than we can handle, but that meaning is corralled into limited and limiting corridors, albeit corridors without end. I doubt a mechanistic biology would consider that the 'function' of an eye is to be seen, to be looked into, no less than to see and to look – as when glances meet.

[7] Darwin himself appears to use the two terms interchangeably in On the Origin of Species. For example, near the end of Chapter XIII, he writes (pp. 454 – 455), "Any change in function, which can be effected by insensibly small steps, is within the power of natural selection; so that an organ rendered, during changed habits of life, useless of injurious for one purpose, might easily be modified and used for another purpose. Or an organ might be retained for one alone of its former functions."

[8] Largely but not wholly unexamined: as physicist David Bohm pointed out, "It does seem odd … that just when physics is … moving away from mechanism, biology and psychology are moving closer to it. If the trend continues … scientists will be regarding living and intelligent beings as mechanical, while they suppose that inanimate matter is too complex and subtle to fit into the limited categories of mechanism." Cited by Carl Woese

meanings, but it has yet to begin to think a biology that does not, in an absolute sense, require a τελος at all.

But now is it a crossroad that we stand at, either to imagine that new biology or to confront head-on why we need a τελος to make sense of the living? It is not. It is but a single path – though passage is far from assured. Already along it, we stand here or there; and pausing for a moment, we imagine once more Menelaus on a beach or, perhaps, in a heath.[9] He pitches his foot against none other than Proteus and grabs on. First he grabs a bearded lion, then in quick succession a serpent, a panther, a torrent of water, a tree with soaring branches, a flame. At last, Proteus turns into a watch. Now Menelaus has him. Now Proteus is caught. All that remains is for him to tell us our future and our end.

It is just possible that, however allegorical this sounds, it is the truth of the world; it is just possible that Proteus' true form – if not his final form or original form – is, wonder of wonders, that most philosophical of devices, the mechanical watch, what human beings use to keep track of time if not of their boredom.[10] But wait! – What are these tracks? Why do we keep them? *How* do we keep them?

And if we lose track, what then?

---

on p. 100 of "Evolving Biological Organization," in Microbial Phylogeny and Evolution, edited by Jan Sapp (2005).

[9] The allusion is to William Paley's Natural Theology (1809), beginning with its very first sentence.

[10] One thinks of the starved tick kept confined in a lab, supposedly without world, for 18 years, as mentioned by Jakob von Uexküll in his introduction to A Foray into the Worlds of Animals and Humans (p. 52). How did the tick pass the time? Uexküll's answer: she didn't! Time stood still! For the tick there was no tock -- undoubtedly because a tick without world has no watch.

II. "For 'tis the sport to have the enginer / Hoist with his own petard..." (*Hamlet* III, iv)

My second thesis is that, as a result of this disruption of kinds, how we imagine those other living beings frequently and perhaps necessarily bleeds into our imagination of ourselves, however much our claims might seem insulated or restricted.[11] This indeed might be the foreseeable consequence of kinship, of that biological blur in which kinds can be declared more easily than found. In lieu of a demonstration, let me give two examples of what I mean.

My first will be taken from Maimonides' The Guide of the Perplexed, a text I choose for the sake of convenience: similar arguments could be made about many other texts in the western tradition, notably those of Descartes, but also, if in different ways, even of Aristotle and Plato.[12] In The Guide,

---

[11] This often comes to mind as I listen to participants in my undergraduate seminars dismiss our 'animal' natures, whose conquest is apparently a not insignificant project of philosophy. One would be wise to question what we mean by "animal" and "animal nature" before vilifying them; wiser still to question why we associate whatever nature this is with animals. What would our treatment of animals be, if we did not conceptualize ourselves as spirits trapped in bodies, if we did not equate spiritualization with loathing of embodiment? With what right, by what perverse logic, in what kind of economy, did we force animals to be mere bodies so that we could uphold our precarious claim on spirit?

[12] I'm thinking, for example, of the manner in which the Cartesian view of animals as automata spreads to human beings, at least when seen through a window; of Aristotle's admittedly problematic distinction of the natural slave (e.g. "Those who are as different [from other men] as the soul from the body or man from beast – and they are in this state if their work is the use of the body, and if this is the best that can come from them – are slaves by nature."

Maimonides argues that providence is consequent upon the overflow of the divine intellect into its human counterpart. It has "singl[ed] out the human species alone among all the species of animals"[13] because, according to Maimonides, humans alone have intellect.[14] He uses this conclusion to justify our treatment of animals, our use of them for labor and for food. It's true that he acknowledges that "[*To avoid causing] suffering to animals is [an injunction to be found] in the Torah*."[15] But even this injunction is for our sake, not the animals': it is to protect us from acquiring moral habits of cruelty where no utility is served. Animals are a means for Maimonides;[16] they are tools, food, things against which to develop moral habits rather than beings with whom to enter into ethical relationships.[17] For Maimonides the τελος of all other living

---

Politics 1254b15 ff, Lord translation); and of the 'brutal' treatment of the black horse in Plato's *Phaedrus* or the interchangeability of human and animal lives in the Myth of Er at the end of *The Republic*.

[13] The Guide of the Perplexed, Part III, Chapter 18, p. 474.

[14] Of course, this claim is worth pausing over. One might wonder how we know it or what we mean by "intellect" for it to be true. For Maimonides as for many others, it is self-evidently true. He considers the question, "Why did He [God] give intellect to man and not to the other species of animals?" (*Op. cit.*, III.18, p. 474) but not *whether* that is so. To the first question he suggests three possible answers: "He willed it so; or, His wisdom has required this to be so; or, nature has required this to be so."

[15] *Op. cit.*, III.17, p. 473. The bracketed additions are in the Pines translation of the text; I did not add them.

[16] I often wonder if this common conclusion doesn't precede its own argument: we treat animals as means, therefore they must truly be means. Because if they were not – !

[17] I am willing to go further. It seems to me that ethical acts are like gifts: you can never be absolutely certain that you've actually witnessed, offered or been the beneficiary of one. Perhaps, when all is said and done, it was only

beings is the τελος humans make of them. This is perhaps not so different from understanding them as watches; if nothing else, in our treatment of other living beings we pass the time.

But where does this human distinctness from animals lead? The answer may be surprising: it leads to the retreat of the distinction. "As for the ignorant and disobedient [human beings]," Maimonides writes, "their state is despicable proportionately to their lack of this overflow [of divine intellect], and they have been relegated to the rank of the individuals of all the other species of animals: *He is like the beasts that speak not*. For this reason it is a light thing to kill them, and has been even enjoined because of its utility. This matter is one of the fundamental principles of the Law."[18]

We should wonder what is fundamental about this principle. We should no less wonder what this utility is in

---

another exchange. Indeed, one common, modern account of animal altruism is specifically that it is veiled reciprocity; and where there is no opportunity for reciprocal exchange, then the animal has simply made a mistake. Such an explanation has, not surprisingly, infiltrated discussion of human ethics: we are generous, we are charitable, as a result of atavistic miscalculations, propensities formerly (but perhaps no longer) adaptive. Notice, however, that the uncertainty that, to me, is the necessary accompaniment of an ethical act or of a gift or of a response has here only been resolved negatively; the act was not ethical, it was not a gift, it was not a response. Freedom reverts back to causality and its cousin, economy. Its incomprehensible *singularity* is tamed into a dull but graspable *kind*. Yet may there not be a world in which the *possibility* of ethical acts, of gifts, of responses, of freedom is respected and preserved, even if not decided? In that world, why would possibilities seemingly so often in evidence among animals be any more easily decided than those among humans – to say nothing of those between animals and humans?

[18]*Op. cit.*, III.18 p. 475.

service of. Is killing ever so light a thing that we shouldn't know its end, its τελος? But *can* we know the end of death?

I also hardly need to point out the political ramifications and appalling history of the thinking which, through recourse to utility, reduces human beings to mere means, indeed to mere animals, as prelude to justifying acts against them that are often, with painful rigor, described as "brutal" and "bestial." Which are the beasts and which the human beings – the perpetrators or the victims?[19] Who's who? Let me add that the tragic 20th century history to which I am in part alluding was often explicitly justified by recourse to Darwinian doctrine. It should confront us with a question of immense and disturbing scope: if we are all somewhat more than kin and less than kind, can we truly think of other living beings as means while preserving for ourselves the category of ends?[20]

This question should also be alive in the perhaps surprising arena of modern education. This is my second example. I'm not the first to suggest that a great deal of

---

[19] I can't help but think of the Sophoclean play *Ajax*. In it Ajax tortures a white-footed ram that, in his divinely wrought madness, he believes to be Odysseus; but Sophocles, ever attentive to the reciprocity of metaphor, also transforms Odysseus into a hound tracking and circling Ajax like prey. I suspect Sophocles was too rigorous to do the one without the other. Shakespeare also comes to mind: "For 'tis the sport to have the enginer / Hoist with his own petard..." Hamlet says in anticipation of blowing Rosencrantz and Guildenstern to the moon – two sentences after first transforming them into adders fanged. (*Hamlet* III, iv) And what of Plato's *Sophist*: is the text not a long hunt for a hunter? (See the second stasimon.)

[20] Are means thinkable without ends, or vice versa? Are relations thinkable with essences, or vice versa? In privileging one or the other of the terms in these pairs, have we done the slightest thing to challenge or displace our metaphysics?

modern education looks like animal training. Everett Dean Martin says so in his 1926 book The Meaning of a Liberal Education.[21] There he calls attention to two criteria that distinguish animal training from a liberal education. First, the liberally educated think of education and of themselves as ends, not as means. Second, they don't simply try to solve an immediate problem but they seek insight into the situation as a whole. The liberally educated seek ends and wholes.[22] Animal training, in contrast, discourages both modes of thinking, both modes of self-consideration. It is limited to questions like, "How do I use this to get that?"; to learning by conditioning rather than by developing insight into a situation; and to following instructions and doing as you're told. According to such a description, I suspect we would have a hard time distinguishing the vast majority of modern educational projects from animal training.

But why do we associate this training with animals? Is it because animals are limited to this type of education? Or is it rather that this type of education is what we enforce upon them, a τελος that we declare for them? Put some apes in a

---

[21] The Meaning of a Liberal Education, chapter 2. Martin goes further than I will discuss here, tracing the tension between liberal education and 'animal training' all the way to ancient Greece, to the philosophical education of the leisure class and the utilitarian education of the slaves. What is it, we might wonder, that for over two millennia – through Scholasticism, Renaissance, Reformation and Industrial Age – has connected these seemingly disparate poles, liberal education and animal training? Is the connection incidental or essential?

[22] Of course, one of my points is that evolution calls both ends and wholes into question. Hence the problem.

cage.[23] Deny them food but place bananas just out of reach. Eventually give them a stick. What questions occur to the apes? The apes who dwell on speculative questions – questions about the justice of the universe, or about the darkness of the human soul, or about their own proper self-conception – starve. The more 'intelligent' apes, however, are driven to instrumental reason: they are the ones who ask, how do I use the stick to get the bananas? In this pedagogical program, only those apes who think of a world of means and ends, who allow themselves to be reduced to mere appetites in need of satisfaction, succeed. In fact, only those apes survive.

I want to emphasize that I have no idea what an ape thinks about in a cage; I wonder about it and I think it's important to wonder about. That some can be coerced into a demonstration of instrumental reason teaches me little about them and more about us. It leads me to worry about what we are doing to ourselves when we reduce education to the approved-of use of instruments which we make little attempt to understand and to the pursuit of jobs whose value is measured in dollars rather than bananas. To my eye, it looks like the distinction in kind which we postulated between ourselves and other living beings has eroded away. A secret kinship, in which neither party is well-served, has instead betrayed the singularity of kinds.[24]

---

23 What follows is heavily indebted to Kafka's "Report to an Academy" and J. M. Coetzee's Elizabeth Costello.

24 This paradoxical phrase – singularity of kinds – is intended to provoke scrutiny. Could there be kinds in a world with neither means nor ends? Could there be speech? Or would there be only writing?

We pretend that this is a world without τελος, that a world without τελος is a world only of means. But in fact this is instead a world of impoverished τελος, a world of τελος violently imposed, a teleological world with an end no one should strive for. Why then are we striving for it? Does being more than kin and less than kind have to lead here?

Or is it rather our metaphysics which leads us here, along this track? – a metaphysics which insists that signification must have a beginning and an end, if only so that passage may be made, bridges built; if only so that our own limits may be surpassed?

Second Episode (ἐπεισόδια): Life of Knowledge – Toward a Post-scientific Post-humanism

Stasimon (στάσιμον): A Sophist

"Well, now it's our job not to let the beast escape (ἀνεῖναι).[1] *We've almost hemmed him in with one of those net-like devices that words provide for things like this."*

The Eleatic stranger in Plato's The Sophist, ~235b.

"What division is more important," the Eleatic stranger asks the young Theaetetus near the end of the dialogue (267b), "than the one between ignorance and knowledge?" "None," answers the mathematician, as if he has already forgotten the difficult work the two of them had undertaken to divide being from non-being. But perhaps he hasn't forgotten. Perhaps instead he sees that the division between ignorance and knowledge is just a veiled form of that other division. As soon as we are confident in the achievement of the one, he thinks, surely we can declare victory over the other. On this view, the old man and the boy confront not multiple problems but a

[1] ἀνεῖναι (aneinai), interpreted here by translator Nicholas White as ἀνα (ana, up) + ἵημι (hiēmi, to send), hence "to send up or forth" or even "to make spring up," whether seeds from the earth or the dead from the grave. It can also mean "to give up" (in the many senses of giving up) or "to let go" including "to let escape." But the word might instead be interpreted as the α-privative coupled to the verb εἶναι (einai), "to be," and our job would then be "not to let the beast not be." Such is the subtle metamorphosis even of a word with unchanging form.

single, if pleomorphic, one – a shape-shifting problem. How strange it would be – but how convenient! – if this protean being (if it is a being) should occasionally transform into some less threatening and easily captured form. If only one! – because then we would grasp it, and would we not as a result have the whole of which it is the partial expression?[2]

For example, what if this being (if it is a being) which endeavors to cross the boundaries between ignorance and knowledge (and being and non-being), should become a minnow or stingray or, quite simply, a fish? Why, we know what an angler is (221b). It is a hunter (like sophists are hunters [222a], like the old man and the boy are also hunters, hunting sophists) – a hunter, however, of fish. Such slippery beasts, fish: the best way to catch them, as anglers well know, is to trick them into eating something. They take something you give them and make it a part of themselves; and now you have them, because you have what you gave them, which in fact was not a gift at all but only the semblance of a gift.[3]

Of course, if you're not an angler but a fisherman of a different kind, you can instead use a net, that weaving together of what is with what is not; and so long as the fish is bigger than the gap, so long as being can't slip through the interstices of non-being (unless it is the other way around) – well, then, you have a good chance of catching him. Even, it turns out, should your net not be a real net but a net of words: words

[2] Thus did all-wise (mētieta, μητίετα) Zeus acquire wisdom, by tricking the Titan Μῆτις (Mētis) into transforming herself into a fly so that he could capture her, eat her and thus internalize her.

[3] Those of us who are fish, then, would wisely beware of what we eat.

which take the place of a net precisely because the net, ironically, isn't there.

To catch the sophist in that less threatening and easily captured form – the form, say, of a fish, rather than the form of non-being – thus requires a net (albeit one that isn't really a net) – or a gift (of the non-gift variety). In the meanwhile, are we not lucky that the hunter we are hunting has so meekly transformed himself, and not us, into the hunted, even if he is still a complex beast which requires more than one hand to handle him (226a)? And what shall we do with our fish once we have him? Dare we eat him?

Would it have been any easier had we thought the sophist became, not a fish, but a god? Or, equivalently, a philosopher? For, like fish, do not philosophers and gods eat? Can they not be contained, if not in nets, then in net-like words? Surely, then, to distinguish a philosopher (who, we know from experience, can also become a sting-ray or a gad-fly) is not harder than to distinguish a god (216c). Gods, like philosophers and fish, can, with some artifice, be grasped and held onto and even eaten – so long as we do our work and not let the beast spring up (whether from seed or the dead, whether from beginning or end); so long as we do our work and not let the beast not be. Our net, or what in any event is like a net, must instead hem the beast in between that springing up and that not-being. The beast must neither come back into existence nor fall back into non-existence. There's our purchase – between the two. There's our hold. That's what we must grasp; grasp, and take within ourselves.

Thankfully, Proteus was no sophist and, though he turned into fire and water and air, he refrained from turning also into nothing. For had he done so, how would Menelaus have grasped him?

Menelaus had no net!

Chapter Three, on Conservation Biology

Grasping Beasts between Being and Non-being: Of Rats and Men

How now! a rat? Dead, for a ducat, dead!
Hamlet, Act III, scene iv

Prefatory note:

This entire chapter is largely anticipated by Nathaniel Hawthorne's "The Birthmark" and its exploration of the grounding of a scientific project in an unexamined or misunderstood aesthetic. Further, a symbol of that aesthetic is nothing other than a grasping hand, in which shape is the titular birthmark found on the left cheek of the character Georgiana. "Had she been less beautiful, – if Envy's self could have found aught else to sneer at, – he [the "man of science," Aylmer] might have felt his affection heightened by the prettiness of this mimic hand, now vaguely portrayed, now lost, now stealing forth again and glimmering to and fro with every pulse of emotion that throbbed within her heart; but seeing her otherwise so perfect, he found this one defect grow more and more intolerable with every moment of their united lives. It was the fatal flaw of humanity which Nature, in one shape or another, stamps ineffaceably on all her productions, either to imply that they are temporary and finite, or that their perfection must be wrought by toil and pain. The crimson hand expressed the ineludible gripe in which mortality clutches the highest and purest of earthly mould, degrading them into kindred with the lowest, and even with the very brutes, like whom their visible frames return to dust. In this manner, selecting it as the symbol of his wife's liability to sin, sorrow, decay, and death, Aylmer's sombre imagination was not long in rendering the birthmark a frightful object, causing him more trouble and horror than ever Georgiana's beauty, whether of soul or sense, had given him delight."

Let me make three observations. First, recourse to the aesthetic is necessary precisely because what is figured is not there. Embodiment, finitude, sin – all of these too are figures, again of what is missing and for which the word "death" would hold catachrestic sway. A metaphysics of presence, insofar as it grounds itself in what it lacks, necessarily aestheticizes. Second, the man of science, in his misguided exertion of supposed mastery over nature, thereby exalts himself into that desired plenitude and eternality – but again only by way of a figure, moreover one unrecognized as such. Aylmer will accordingly dedicate his seeming knowledge to the aesthetic project of removing his wife's birthmark – and, unbeknownst to him but foreseeably, ending her life. In the process, he inscribes himself as another version of the same threatening hand which he reads the birthmark as representing. Finally, one should note that the birthmark is there, as it were, from the very beginning: "Do we know that there is a possibility, on any terms, of unclasping the firm gripe of this little hand which was laid upon me before I came into the world?" asks Georgiana, like her husband not realizing the import of "on any terms" when at stake is getting rid of "this fatal birthmark" – the one which grips us all. In short, Hawthorne's tale preempts many of this chapter's observations and much of its argument.

I. Introduction: Rewilding the Galápagos?

In a short commentary in *Nature* in 2005, and in considerably more depth in a 2006 issue of *The American Naturalist*, Josh Donlan and 11 colleagues offered what they described as a bold plan to reconstitute conservation biology's underlying premise "from managing extinction to actively restoring natural processes" (Donlan et al. 2005). Specifically they proposed rewilding North America with Pleistocene megafauna, the extinction of which 13,000 years ago was almost certainly due in part to human actions. Because of that complicity, and because "our subsequent activities have curtailed the evolutionary potential of most remaining large vertebrates," Donlan et al. (2005) asserted that we "bear an ethical responsibility to redress these problems." Without lingering over the serious question of what kind of redress, if any, is possible, they instead proposed introducing related megafauna – horses, camelids, cheetahs, African and Asian elephants, and even lions – as proxies for their extinguished North American cousins. The "aesthetic and economic benefits" of such a plan are "clear," they claimed, citing their Figure 1, a cartoon pretending to quantify such elusive categories as "potential economic/cultural value" and "potential economic/cultural conflict." (Their longer 2006 paper offered little elucidation of how such categories are to be judged, much less quantified.) As further benefits, they noted that North America would become an additional refugium for endangered megafauna from other continents, while their reintroduction would fuel an ecotourism boom ultimately expressive of "our own Pleistocene heritage" and the "emotional relationships with large vertebrates" then engendered – even though those relationships culminated in the very extinctions for which this project would atone.

Of course such a proposal stimulated both interest and controversy. New York Times columnist Nicholas Kristof was all for it, as a way to entice Americans to "connect... with nature" (Kristof 2005). Tim Caro (2007) catalogued many serious concerns about Pleistocene rewilding but concluded generously that it deserved merit as a "proactive idea that could galvanize the conservation community out of its helplessness." Others were more incredulous and critical. Not only have North American ecosystems changed in the intervening 13,000 years, but introduced species typically cause loss of biodiversity, noted Smith (2005). Oliveira-Santos and Fernandez (2010) argued similarly and added the sobering observation that the ecological roles of introduced organisms are notoriously unpredictable. It's not possible to restore the evolutionary potential of extinct species, Rubenstein et al. (2006) pointed out; introducing proxies is not the restoration but the invention of an ecosystem. Why not instead try to restore the historical ranges of extant species like grizzly bears and pumas, rather than introduce lions and cheetahs? Others challenged as too simple or optimistic Donlan et al.'s anticipation of the ecological, economic and cultural impediments (Dinerstein and Irvin 2005; Schlaepfer 2005; Shay 2005).

My own interest was stimulated by a serendipitous observation. Virtually simultaneous with the Pleistocene re-wilding proposal was an ongoing and historically ambitious campaign to exterminate organisms introduced to the Galápagos Islands by human beings (Kaiser 2001; Guo 2006; Lavoie et al. 2007). Attempted, succeeding and/or successful eradications on individual islands were documented, often in miniscule detail, for goats (Campbell et al. 2004), pigs (Cruz et al. 2005), donkeys (Carrion et al. 2007), black rats (Appendix 1 in Howald et al. 2007), and the little fire ant *Wasmannia auropunctata* (Causton et al. 2005). These eradication attempts were frequently on an unprecedented geographic scale, but more striking to me was the toll, the body count beyond or nearly beyond my imagination, and the meticulous documentation of extermination often down to the very last animal: 1,523 donkeys, > 18,800 pigs, 201,825 goats (including > 140,000 during a single campaign called "Project Isabela"; Carrion et al. 2011), and ~ 180 million black rats as part of "Project Pinzón" (Associated Press 2012; Nicholls 2013). What further caught my eye was the involvement of Josh Donlan, who is listed as corresponding author on all of the scientific papers documenting vertebrate exterminations cited above. Rather cold-bloodedly, I found myself wondering at a logic that proposed in one place widespread introduction of megafauna to compensate for Pleistocene extinctions partially attributable to human hunting, and yet, in another place, hunted introduced species, ironically including megafauna, to local extinction – all the while explicitly justifying both projects as restorations grounded in knowledge (Donlan et al. 2005; Carrion et al. 2011).

This chapter was born out of the seeming tension and yet interconnection of those two projects. Both take for granted a familiar if still paradoxical understanding of the role of human beings in nature, as well as the propriety of a presumed knowledge arbitrating that role. Whether in Pleistocene North America or the modern-day Galápagos, human presence and activities (hunting animals to extinction in one place, introducing organisms in the other) are identified as a necessarily corrupting influence which, curiously, only human presence and activities, but now supplemented by knowledge, can possibly redeem, albeit through precisely the same gestures as formerly were corrupting – introducing organisms in one place, hunting animals to (local) extinction in the other. Such an understanding extends beyond Donlan and his colleagues and is in fact both pervasive and widely accepted in conservation biology, as exemplified in the recent review by Seddon et al. (2014; e.g., their Figure 3). But the question is, how exactly are we to distinguish our roles as corrupters from our roles as redeemers? Where in fact does corruption end and redemption begin? Is it really knowledge which differentiates them? Finally, how coherent is this understanding of the human role, or rather, more importantly, what are the conditions of its coherence?

To approach these questions, I explore the justifications put forward for both projects, under the headings by which Donlan et al. (2006) organized their rewilding case: "ecological arguments," "evolutionary and conservation benefits," and "cultural and economic benefits." The earlier 2005 paper expressed itself in similar terms, founding its proposal on "ecological, evolutionary, economic, aesthetic and ethical grounds." But what precisely are those grounds? Can any of them provide the necessary distinguishing criteria between human redemption and human corruption? In short, on what basis, ultimately, are we to adjudicate our responsibilities to mastodons, cheetahs, tortoises, goats, rats, ants and even ecosystems, past, present and future? It will turn out that the authors' answers to these questions always invoke knowledge, moreover a knowledge that is seemingly unconditional, acontextual and therefore, presumptively, universal. Understood in this way, knowledge, whether real or imagined, serves as both unquestionable foundation and as alibi, relieving us of the burden of introspection, doubt and decision. In parallel, the question of how to adjudicate our responsibilities is not truly answered by knowledge so much as evaded by its invocation, much like the related question about whether redress is possible in the first place.

I argue that this evasion, and the theory of knowledge of which it is a symptom, is ultimately an expression of implicit metaphysical commitments that privilege presence, without noticing that "presence" is only ever brought into the world as a simulacrum. I show how the scientific arguments – the arguments based in science, in *scientia*, in knowledge – for these projects are both inconsistent on their own terms and an expression of those commitments. Likewise the more humanistic arguments. The critique that I am offering instead would cease to evade not only important but impossible questions and instead ask them – that is, endure them.

II. The Scientific Case: Ecological Arguments

"We can no more predict the future composition of communities than some Ordovician ecologist could have foreseen the Great Barrier Reef." (Myers and Knoll 2001)

Donlan et al. (2006) presented three ecological arguments in support of rewilding North America, as follows:

(1) They inferred both the likely importance of megafauna in Pleistocene ecology and the likely ecological consequences of their extinction, some of which remain in evidence to this day (for example, in the disappearance of most large-fruited *Maclura* and in the arguably "overbuilt" speed of the pronghorn antelope). If the loss of apex predators like gray wolves led to an increase in herbivory, consequent shifts in the abundance and distribution of specific plants, and cascading effects that ultimately included sediment and nutrient dynamics – all of which the reintroduction of wolves to Yellowstone suggests – how could the loss of vertebrate megafauna not have had similarly dramatic and far-reaching consequences?

(2) They suggested likely ecological benefits to what they termed "reintroduction." Continuing the wolf example, they argued that, following the extinction of gray wolves in eastern North America, deer proliferated and helped support larger black-legged tick populations which in turn furthered the spread of the Lyme-disease-causing bacterium *Borrelia*. "[T]he extinction of gray wolves from eastern North America

therefore was plausibly an essential ingredient in the recent Lyme disease epidemic" which might be attenuated "through reestablishment of gray wolves in that region."[1] They further cited many rodent-transmitted human diseases and suggested that reintroduction of large predators might strongly influence their prevalence.

(3) They noted the valuable experimental opportunities that such introductions represent, for example, to better understand top-down regulation and food web dynamics over large spatial scales.

I concede that the loss of Pleistocene vertebrates must have affected biodiversity and ecosystem function, with far-ranging implications, some of which undoubtedly have been deleterious – to someone. And beneficial – to others. Further, I concede that the experimental opportunities are also great – although those opportunities would no less pertain to the North American introduction of endangered Tasmanian devils. What is not obvious to me, however, is how these arguments facilitate the case for rewilding, on two grounds. First, the very difficulty of foreseeing consequences should give us pause, as noted routinely (e.g., Smith 2005, Oliveira-Santos and Fernandez 2010 and Seddon et al. 2014). For example, the introduction of large carnivores might suppress intermediate predators and lead to a proliferation of plague-transmitting

---

1 Analogous arguments have been made to explain the Lyme disease "epidemic" as a consequence of the extinction of the passenger pigeon (Blockstein 1998). As frequently if not always with evolutionary arguments, the problem is how to distinguish what is plausible from what is true.

rodents – hardly a benefit at all, from the human perspective. But that perspective is the second problem: on what *ecological* grounds do we privilege it? While Donlan et al. (2006) described present-day, megafauna-impoverished ecosystems in negative and often morally valenced ways (they discussed them in terms of "disequilibrium," "degraded systems," "ecosystem dysfunction" and cite a paper on "ecosystem decay"), it is unclear what justifies those judgments. Ecosystems may be more or less resilient, efficient, biodiverse and interesting to us – but ecology is in no position to valorize any of those states except in terms of their meaning *to us*, which is to say, in terms of some ancillary, non-ecological criterion. But if this is the case, then the ultimate justification for rewilding must be found elsewhere than ecology. It also problematizes a campaign born in reaction to supposed human interference in nature, but which privileges certain ecological states over others precisely because of their concordance with human aims and desires.

These critiques find expression, curiously enough, in another article co-written by Donlan (Roemer, Donlan and Courchamp 2002). There the authors pointed out the dramatic, unforeseen consequences of the introduction of pigs to an island ecosystem, culminating in the "wholesale restructuring of [the island's] food web" due to the now-possible colonization of a new island predator (golden eagles), the resultant near extinction of the endemic island fox and the subsequent release of populations of a "competitively inferior" island spotted skunk. The reason that these effects were not foreseen is also the reason why I have felt compelled to put "competitively inferior" in quotation marks: we are regularly tempted to see as essential features of an organism or an ecosystem characteristics which are instead relational, hence contingent, hence (often) unpredictable. The skunk was competitively inferior only circumstantially; when those circumstances changed, it ceased to be so; and that now-recognized relational characteristic was, unsurprisingly, not predicted. Similarly, there is no essential Pleistocene ecosystem that can be recovered; what existed in the Pleistocene was a dynamic system of relations, only some of which can be reconstituted – *in approximation only, metaphorically, rhetorically, through proxies* – after 13,000 years of change. The problem with North American restoration is that there is nothing, no *thing*, to restore. Even Donlan et al. (2006) admitted this, through their care to emphasize restoration, not just of species, but of "species interactions" – "the true functional fabric of nature." But they seem not to have deeply considered how interactions – "the mutual relations of all organic beings," our ignorance of which Darwin considered a "necessary conviction" in Chapter III of On the Origin of Species[2] – can be rediscovered, much less imposed, across an immense gulf of time and change. Nor do they appear to consider very seriously the enormous expansion and transformation of the human role since the Pleistocene, a role which would unavoidably play a dramatic part in any "reconstituted" ecosystem.

Similar justifications are implicit to the ecological argument for restoring the Galápagos – and a similar critique pertains. That the ecology of these islands has been radically altered since the first recorded human discovery of them in the

---

[2] All references to On the Origin of Species (henceforth simply called The Origin) will be to the facsimile of the first edition.

16th century (perhaps although controversially preceded by pre-Columbian visitors; Flett and Haberle 2008) cannot be disputed. Seventeen recently-introduced organisms were observed by Darwin in 1835; now there are > 900 introduced plant species alone (Vince 2011). "What we're doing is homogenizing the world's biota; setting the world on a geological epoch: the Homogocene," moaned the ecologist William Laurance specifically with respect to Galápagos conservation (as quoted by Vince 2011). Numerous endemic species have gone extinct while, and often because, faunal and floral immigrants brought by human beings, whether deliberately or accidentally, have flourished. The myriad relations which constitute an ecosystem have undoubtedly changed, with often major biological and physical ramifications – as dramatic photographs of goat-grazed and goat-protected enclosures on the islands make clear (e.g., Lavoie et al. 2007, p. 24). Worse, the losers tend to be charismatic, like the giant tortoises, while the winners are organisms whom we habitually disparage: "pests and weeds" (Donlan et al. 2005), "junk species" (Vince 2011) and "tramp species" (Clark 1980, Pezzatti et al. 1998), not to mention the common terminology of "aliens" and "invaders" (e.g. Kaiser 2001).

Colorful – and disturbing – epithets aside, the problem of privilege remains: which species, which relations, which ecosystems should we privilege, and why? This problem is magnified on geologically young islands like the Galápagos, all of whose denizens are, on the scale of geological time, recent immigrants. No doubt for this reason, Darwin, in his rare mentions of the Galápagos in The Origin, routinely referred to their myriad biological inhabitants as "immigrants" or "colonists" (see, e.g., Chapter XII, pp. 397 – 403). So who really counts as alien, who as native? Who corrupts and who redeems? "Cultural conventions dictate which taxa are regarded as native and which are not, usually irrespective of ecological and historical insights," Donlan et al. (2006) wrote. But do ecological and historical insights manage to obviate the dictates of cultural convention? Do they truly attain neutral universality?

Consider *Rattus rattus*, the Old World black or ship rat. It has been present in the Galápagos almost certainly since the late 17th-century visits of buccaneers (Patton et al. 1975). It is considered an introduced species. But the endemic Galápagos rat which Darwin observed on San Cristóbal Island in 1835, classified by Waterhouse shortly thereafter as *Mus galapagoensis* and later reclassified in 1892 as *Oryzomys galapagoensis* (now considered extinct but a conspecific of the extant *O. bauri* on Santa Fe island), is probably a Peruvian immigrant arrived in the last few hundred to a thousand years, brought "perhaps by aboriginal sailors coursing the west coast of Peru" (Patton and Hafner 1983). Darwin suspected as much: "Although no one has a right to speculate without distinct facts," he wrote in The Voyage of the Beagle (1962, p. 379), "yet even with respect to the Chatham Island [= San Cristóbal] mouse,[3] it should be borne in mind, that it may possibly be an American species imported here; for I have seen, in a most unfrequented part of the Pampas, a native mouse living in the roof of a newly built hovel, and therefore its transportation in a vessel is not

---

[3] Here the reference is clearly to *Mus galapagoensis*.

improbable…" Nonetheless, despite their likely historic colonization with possible abetment by man, *Oryzomys* is native. Indeed, black rats must be extirpated in part because of the apparent havoc that they have wreaked and will continue to wreak on indigenous rats (e.g., Brosset 1963).

Yet, in other cases, the mere suspicion of human agency in an organism's arrival at the islands justifies dramatic action against it. Even birds capable of flight, including smooth-billed anis and rock doves, are classified as introduced. In the case of the ani, Rosenberg et al. (1990) and Grant and de Vries (1993) made acknowledgedly circumstantial cases that it did not arrive naturally, because a weak flyer. (Kramer [1984] drew the opposite conclusion, considering anis independent arrivals capable of settling, however, only because of human modification of the islands.) Making a yet-more-circumstantial case that they are harmful to indigenous Galápagos organisms, Rosenberg et al. (1990) did not hesitate to urge their extermination. They are now commonly listed among removal targets (e.g. Kaiser 2001, Lavoie et al. 2007). It is correspondingly ironic that the poster child for the great success of the eradication campaigns against goats and pigs is the resurgent Galápagos rail (Donlan et al. 2007, Lavoie et al. 2007, Carrion et al. 2011) – a similarly weak-flying bird, but one whose native status has yet to be questioned.

Besides uncertainty (as with the ani) and/or inconsistency (as with *Oryzomys*) about the role of human-mediated transfer of species, two additional issues arise, one theoretical and one practical. The theoretical issue is this: the annals of natural history are rich in examples (including from the Galápagos; Clark and Clark 1981) of elegant, sometimes elaborate adaptations that organisms have developed to facilitate their dispersal by other organisms. No one disputes this. Indeed, Donlan et al. (2006) lamented that "[t]he loss of proboscideans and other megaherbivores capable and suspected of dispersing the large fruits of these trees may have caused or contributed to the extinction of the other *Maclura* species." (They continue: "...whereas Osage orange [the only extant *Maclura* species] fortuitously survived as a small remnant and spread because of dispersal by modern humans.") Darwin conducted numerous experiments, many of which are described in The Origin, on means of dispersal, including by birds, mammals, fish and insects, whether living or dead, and called repeated attention to adaptations of one organism to dispersal by another. "[F]ew relations are more striking[4] than the adaptation of hooked seeds for transportal by the wool and fur of quadrupeds," he wrote in Chapter XII (p. 392). So what is the essential difference which would distinguish an organism adapted to dispersal by elephants, migrating birds, fish, insects or furry quadrupeds from one adapted to dispersal by humans? Is only the latter organism cheating?

The practical issue boils down to a simple question: can anyone seriously believe that human-introduced species in the Galápagos can be controlled so long as ~180,000 tourists visit per year (Hoyman and McCall [2013] and compared to ~100,000 reported by Boersma et al. [2005] and 12,000 in 1979 [de Groot 1983]), a fishery has blossomed into an industry serving international markets (Boersma et al. 2005), and ~30,000 people have settled (with the population growing at a

[4] In subsequent editions, Darwin considered this relation not striking but manifest.

6.4% annual clip; Nash 2008), many to support the tourists (and not a few to support the resident scientists)? In 2008, the Galápagos were described as "one of the fastest growing economies on the planet" (Nash 2008), a circumstance whose major driver was ecotourism. Indeed, the Galápagos are the object of numerous, romantically-named ship tours (e.g., Wildland Adventures, Lindblad Expeditions, Ocean Adventures, Ecoventura – not to mention Adventures by Disney); the subject of no less numerous romanticizing films (including a 1934 silent film "The Empress of Floreana," two IMAX films, a BBC documentary series, and a sequence in the 2003 blockbuster "Master and Commander"); and are featured regularly in popular magazines like *National Geographic* and *National Geographic Traveler*. They are the province of a notoriously unstable government which lacks the resources and often the will to enforce its own ambivalent decrees about the islands. Meanwhile, even the Charles Darwin Research Station (CDRS), founded with (and operated by) the Charles Darwin Foundation (CDF) in 1959 in part to eliminate introduced species (de Groot 1983), but also tied to the promotion of tourism since the United Nations Educational, Scientific and Cultural Organization (UNESCO) reconnaissance mission in 1957 (Cairns 2011, p. 81),[5] has expanded to

[5] In Barrow's 1960 report on the 1957 UNESCO reconnaissance mission to the Galápagos, he both noted the dangers of tourism (calling "transformations in the local scenery and biota" concomitant with tourist access "inevitable" [p. 38]) while on the same page ambivalently observing that only the tourist with "a naturalist's heart" could appreciate the islands. He went on to recommend the founding of a research station on the islands, "one of the purposes [of which would be] to encourage scientific investigations in this most interesting area, and also to preserve as much of

include satellite offices on Isabela and San Cristóbal besides its main office on Santa Cruz. "A question often arises to what degree the [conservation] endeavor is quixotic," Loope et al. (1988) somewhat understatedly confessed. Boersma et al. (2005) were less sanguine: simultaneously promoting sustainable development (much of it through tourism) and biodiversity conservation (also frequently if problematically linked to tourism) on the Galápagos is "a dream uniting incompatible goals." But what kind of dream is it, precisely, to purge the islands of particular human influences – easily visible ones like goats, pigs and rats (but certainly not the no-less-visible but rarely photographed tourists) – while subjecting them to a historically unprecedented degree of human visitation? Anthropology knows that every observer is a participant; physicists have long reconciled themselves to inescapable observer effects; but conservation biology would be different?

So long as human beings are visiting or residing on the islands, as well as impacting the planet globally (e.g., through climate change and ocean acidification), they represent one of the many wild cards the consequences of which can be foreseen only partially and often only in broad terms. Unintended effects are inevitable. That they will occur is

---

the biota as possible in its natural state for future generations to enjoy" (p. 43). Either he imagined that future enjoyment limited to scientists whose professional role precluded their simultaneous identification as tourists, or he was anticipating some form of future tourism explicitly linked to science. That linkage was to a certain degree already implied when he listed among the advantages of Santa Cruz island as the location for the eventual research station Ecuador's plans to construct a road there (p. 52), though he had cited those same plans on p. 38 as being explicitly in the context of developing tourism.

predictable; what they will be frequently is not. The 2001 Galápagos oil spill by the *Jessica*, a tanker there to refuel a cruise ship, is an example *par excellence*. Since then, the number of annual tourists has almost doubled; who can reasonably expect that similar events will not occur in the future?

Yet the deliberate ecological manipulations in the Galápagos fail to take seriously the necessity of accident and the inadequacy of knowledge. Already the elimination of (recently) introduced organisms has led to undesired consequences, including greater proliferation of other introduced species like guava (*Psidium guajava*; Eckhardt 1972, Zavaleta et al. 2001) and blackberry (*Rubus niveus*; Carrion et al. 2011). Likewise, the extirpation of *R. rattus* is likely to lead to an explosion of the similarly introduced *Mus musculus* (Harris and Macdonald 2007, Harper and Cabrera 2010) and, so long as human beings are routinely visiting the islands, will always be temporary. Elimination campaigns have also made use of dangerous chemicals famous elsewhere for their collateral damage, including the rat poison brodifacoum,[6] the insecticide DDT and the endocrine disruptor methoprene (Wetterer and Porter 2003). The effects of these on non-target organisms in the Galápagos are seldom specifically known and, because of limited funds and personnel, have only occasionally been investigated experimentally or in the field (Abedrabbo 1994). Anecdotal or un-peer-reviewed evidence indicates that rat poisoning is associated with at least short-term declines of

---

[6] 4.5 metric tons of which were required for Project Pinzón (Nicholls 2013). A Wisconsin company donated the poison, "attracted by the philanthropic cause."

Galápagos hawks and short-eared owls (Harper and Carrion 2011). For the massive Project Pinzón rat-poisoning, the island's Galápagos hawks were actually held in captivity for 6 weeks after the bait drop; upon release, 16 of the 60 (27%) birds promptly died, presumably from eating poisoned prey (Nicholls 2013). Similarly, one of the few quantitative studies on collateral damage estimated overall mortality of local land iguanas due to brodifacoum at 4.5% of the population, but possibly as high as 10% (Harper et al. 2011).

Meanwhile, in the fight against introduced species, the logic comes full circle and new species are deliberately introduced (e.g., an Australian ladybug to control the cottony cushion scale [Calderón Alvarez et al. 2012], described as a "major threat to both rare endemic species and" – of course without noting the irony – "to agriculture";[7] Simberloff et al. 2011) or are proposed for introduction to serve as biocontrols (e.g., of *P. guajava*; Loope et al. 1988). This despite the salutary lessons of past deliberate introductions, such as the devastation (including extinction) wreaked against > 100 endemic species on numerous Pacific islands when the snail *Euglandina rosea* was deployed as a biocontrol of the giant African snail *Achatina fulica* (Hadfield and Mountain 1981, Clarke et al. 1984). In this context, it is also amusing to note that part of the circumstantial case against the smooth-billed ani as human-

[7] Similarly, in Bowman's 1960 report on the UNESCO reconnaissance mission to the Galápagos, he called attention specifically to the deleterious effects of introduced mice and rats on island agriculture, including (introduced) crops like pineapple, bananas, citrus fruit, potatoes and corn (p. 36). As he envisioned it, the island research station would also include agriculture among its research mandates (p. 55).

introduced is the hypothesis of a motive: farmers, it is said, probably introduced the ani under the misguided notion that it could control cattle ticks (Rosenberg et al. 1990). Apparently the farmers and the scientists share the same premises, but in service of different – but not radically different – goals.

The point is that the sequelae of attempted Galápagos restoration as of Pleistocene rewilding are not wholly predictable, even when their advocates promise (as they always do) "sound research and prescient management plans" and insist that the projects be "science-driven" (Donlan et al. 2006). No research is sound enough, no management plan sufficiently prescient. (Indeed, what is the appropriate time horizon of prescience?) There is a necessity to accident which does not, however, make the specific accidents themselves foreseeable. Further, since accidents are defined with respect to human intention, the more human beings involved in these projects, the more intentions and the greater the probability of accident. This greater probability, indeed this necessity, of accident reflects ways in which our intentions are not actually known to us – ways in which they fail to be fully present to us, although we always assume, without the slightest evidence, that they could be. Hence, without going into a typology of accidents, one can at least observe, as Freud did, that accidents often result from intentions in unnoticed conflict with one another. What better example of such a conflict than the contradictory roles given human beings in both the rewilding and the Galápagos dramas? We corrupt and we restore; a "natural" state defined by human absence is to be reconstituted through human presence. An accident will always be that circumstance when we thought we were restoring what instead we were corrupting.

Rather than trying to referee the human role with an inhuman vigilance – which, besides assuming through a temporal unfolding that we are only a corruptor or a redeemer and never both at once, simply reaffirms a metaphysics which insists on our capacity to be fully present and hence infinitely watchful – it would be better to question the very distinction between the human and the natural being presupposed. We could instead recognize "the reality of human-dominated ecosystems, rather than the separation of humanity and nature," as Western (2001) advocated; or, falling short of that, at least we could recognize, as a few have (e.g., Hobbs et al. 2006, Vince 2011), the potential value of so-called "hybrid," "novel" or "emerging" ecosystems which include introduced organisms – no less than of "restored" ecosystems, themselves (unconfessedly) hybrid, novel or emerging, but nonetheless aspiring to an impossible and surely mythical purity. Best, however, would be to challenge the metaphysics which frames these problems and their solutions in terms of intention, knowledge and presence – thereby establishing the forever unstable opposition between the human (understood as intentional) and the natural (understood as nonintentional) in the first place. Perhaps then we could dispense with the sublime naïveté which characterizes our perpetual surprise when an accident has once again revealed our necessary confusion of myth and reality.

III. The Scientific Case: Evolutionary and Conservation Benefits

"Judging from the past, we may safely infer that not one living species will transmit its unaltered likeness to a distant futurity." (Darwin, The Origin, Chapter XIV, p. 489)

The argument for rewilding based on evolutionary and conservation benefits boils down to the preservation of evolutionary potential: "[B]old actions [are] needed to preserve evolutionary potential in the wake of global change and the drastic decline in biodiversity" (Donlan et al. 2006); "North American sites could serve as additional refugia to help preserve that evolutionary potential [of Earth's remaining and endangered megafauna]" (*Ibid.*); "Here we outline a bold plan ... to restore some of the evolutionary and ecological potential that was lost 13,000 years ago" (Donlan et al. 2005); "[O]ur subsequent activities [since the Pleistocene] have curtailed the evolutionary potential of most remaining large vertebrates" (*Ibid.*); etc. On the one hand, who can dispute this? Of course the Pleistocene extinctions, as well as the ongoing mass extinction due in large part to human activity (Dirzo et al. 2014), curtailed evolutionary potential and will continue to do so, as Donlan et al. claimed in both articles. But what is extinction if not that, the curtailment of evolutionary potential? How does evolution proceed except through the (often selective) curtailment of potential? On the other hand, as a consequence of those extinctions, other evolutionary potentials were and are being realized. For example, the Paleogene diversification of mammals is not unrelated to the

K-T extinction (of, e.g., dinosaurs) which preceded and probably in part enabled it. Likewise it is hard to imagine that the Cambrian explosion is unrelated to the massive biological attrition associated with the prior global or near-global Neoproterozoic glaciations (Hoffman et al. 1998). So the question now becomes, does evolution provide us with criteria by which to distinguish the potentials we should favor?

Narrowly speaking, it appears to. We should favor potentials conducive to our own continuation – assuming, *concesso non dato,* that this is really a matter of human insight and conscious choice. If this is the case, however, then it isn't really evolution which justifies these projects so much as naked self-interest.[8] The argument would then need to be made that these actions are *in fact* in our self-interest, understood in the rather impoverished terms of survival and reproductive success and evaluated over some time period. Of course, that temporal evaluation is a major problem: exterminating Pleistocene megafauna may very well have once been in the (short-term) interest of human beings, even if arguably it was not in our (long-term) interest. So too the ongoing mass extinction. What is the appropriate timeframe over which to assess an

---

[8] One observes that, on such terms, it is no longer clear that rewilding and restoration could continue to be justified on ethical as well as evolutionary grounds. Self-interest, however conceived, may very well justify the sorts of eradications for which rewilding would atone; indeed, on the Galápagos, who could be so confident as to declare that the eradications there undertaken and contemplated are not done in part out of self-interest? Does the tourism company Lindblad Expeditions provide funding for massive extermination campaigns (as it does) only out of disinterested dedication to conservation? Are the scientists involved in no way motivated by their own career goals?

(attempted) evolutionary strategy? The future lasts a long time. It is precisely this problem which regularly besets conservation biology, when stakeholders adopt opposing goals because acting within different temporal horizons: e.g. the hunter of bushmeat feeding his family tonight versus the conservationist trying to preserve a species or ecosystem process for later. Thus it is not clear that even the evolutionary criterion of self-interest truly provides us with the necessary means by which to distinguish the potentials we should favor. But beyond that narrow criterion, into which we may have extremely limited insight anyway, evolution disturbingly provides no other. It offers no value judgment. Were we to bequeath a world bereft of megafauna (including us) to cockroaches, or were the world to abandon, partly through our fault, its metazoan experiment altogether and revert to the microbial predominance characteristic of most of life's known history, evolution (dangerously personified) will not chastise us nor so much as opine. *Cui bono* is our question, not its.

In this context it is important to notice the immense tension that ought to exist between an evolutionary justification and one based on the human project of conservation. Evolution is seldom, and ultimately never, preservative. Exceptions like so-called "living fossils" prove the rule.[9] Instead, evolution is a tale of enforced change and of loss, of seeming ruthlessness often exercised especially against the young, the old and the weak. Even that "principle of preservation" which Darwin, for the sake of brevity, called natural selection, paradoxically but necessarily entails extinc-

[9] For an important caveat against the quick acceptance of "living fossils," see Chapter 1, footnote 36.

tion. Darwin no less than the rest of us struggled immensely with this problem in The Origin.[10] In the last sentence of Chapter III (p. 79), reflecting on the starkness and apparent cruelty of the struggle for existence, he offered that "we may console ourselves with the full belief, that the war of nature is not incessant, that no fear is felt, that death is generally prompt, and that the vigorous, the healthy and the happy survive and multiply." Such may be our consolation – but that does not make it true. In the next chapter (p. 83), he emphasized that Nature selects "only for [the good] of the being which she tends," but by the end of The Origin, we learn that that good does not entail the endurance of any species into a distant futurity. So what being, precisely, will have been tended? What kind of tending will it have been? As argued especially in Chapter 2, few ideas critique the human privileging of presence more than evolution: there is a reason why we have to evaluate our intended evolutionary "strategies" against an indefinitely receding temporal horizon, just as there is a reason why we can frame evolution only in terms of imaginary abstractions like "species." What is "present," evolution always shows, is never sufficiently present to stay present.

Further, attempts to forestall evolution are always simply entrained into the very evolutionary dynamic that they would oppose. For example, conservation projects are heavily biased toward "charismatic" organisms, even while we seldom have deep insight into *why* they are charismatic *for us* – just as a bird may lack insight into why she prefers brightly colored plumage

---

10 This is further discussed in Chapter 2 particularly in the context of Darwin's repeated qualification of extinction as *almost* inevitable.

in a mate.[11] Hence more amphibians are classified as threatened than mammals, reptiles or birds; and in terms of rates of decline, range contraction and species extinction, terrestrial invertebrates are suffering at least as severely as vertebrates and probably more so (Dirzo et al. 2014). Yet,

"…[C]onservation translocation projects, as with other types of conservation management, show a marked taxonomic bias towards birds (33% of projects, whereas birds make up 18% of species represented in nature) and mammals (41% of projects versus 8% of species), particularly the larger, more charismatic species, almost irrespective of the degree of threat or vulnerability. Recent data on reinforcements show that this bias towards birds and mammals is continuing. For conservation translocations in general, relatively few invertebrate, reptile, amphibian or fish species are represented relative to their prevalence in nature" (Seddon et al. 2014).[12]

---

[11] Who is to say, however, what insight a bird has into her preferences? Surely we ought to be amused when Darwin opines that "we see peculiarities arising and becoming attached to the male sex in our domestic animals... which we cannot believe to be useful to the males in battle, or attractive to the females" (Chapter IV, p. 90). He goes on to give the example of "the tuft of hair on the breast of the turkey-cock, which can hardly be either useful or ornamental to this bird..." Why would a turkey wear a tie, after all? Darwin's pairing of the useful with the ornamental is, incidentally, suggestive; and we might accordingly wonder, alongside the Kant of the third Critique, how related judgments of utility are to aesthetic judgments.

[12] I will pass over in silence the myriad difficulties of assessing relative abundance "in nature", both because the relative abundance *of what* is always a question (as discussed in terms of the species concept in Chapter 2) and because the meaning of "in nature" turns out to be far from evident (see below). Suffice it to observe that here I am endeavoring to offer a criticism of

In short, our conservation practices are *selective* – in large part unconsciously – and accordingly contribute to an evolutionary process of change that ostensibly they would abate. Insofar as at least some of the criteria of selection are obscure and undoubtedly entail what may loosely be termed cultural considerations, the so-called "evolutionary and conservation benefits" of projects like re-wilding ultimately rely on external criteria for their justification, as we saw the ecological argument did as well.[13] To recognize this is hardly revolutionary; in fact, it has long been acknowledged, e.g. by the former head of the Natural Environment Research Council

---

conservation practices in their own terms, terms which the field, however problematically, accepts as meaningful.

[13] On this score, it is worth considering the arguments made by Sherkow and Greely (2013, p. 33) in favor of de-extinction projects (discussed in this book's Parodos and clearly closely related to re-wilding): "The last benefit [of de-extinction] might be called 'wonder,' or, more colloquially, 'coolness.' This may be the biggest attraction, and possibly the biggest benefit, of de-extinction. It would surely be very cool to see a living wooly mammoth. And while this is rarely viewed as a substantial benefit, much of what we do as individuals – even many aspects of science – we do because it's 'cool." Greely has reiterated this position elsewhere (as quoted by Zimmer 2013): "'What intrigues me is just that it's really cool,' Greely says. 'A saber-toothed cat? It would be neat to see one of those." There may be more promise in such an acknowledgement than at first appears and certainly more than these authors investigate. What is coolness? What makes projects like these cool? Are they cool in the same way superheroes and superhero movies are cool? I am not at all dismissive of "coolness" as a reason underlying de-extinction or re-wilding – it may very well be, in a certain sense, such a reason – but its bald declaration should not be the end of inquiry (as Sherkow and Greely treat it), but its beginning. We should also draw the obvious conclusion that coolness is not an intrinsic trait of, e.g., wooly mammoths.

(NERC) and trustee of the World Wide Fund for Nature (WWF) UK, John Lawton (1997): "[Conservation] decisions are ultimately political, ethical, aesthetic, even religious, and embrace much more than scientific information." Indeed, he added: "at its heart, setting management objectives for conservation is not a scientific activity."

And yet science is commonly invoked nonetheless. Just as we seek to impose the enduring presence of specific organisms on an evolutionary scene constantly sweeping them away, so do we insist that our imposition is justified by knowledge. Nowhere is this more true than in the Galápagos, where it is implicit in the very rhetoric of description: the islands are extolled routinely as, e.g., an "unparalleled scientific laboratory" (Eckhardt 1972), "a natural laboratory" (Benchley 1999), both a "museum of evolution" and "an important laboratory ... of a truly Darwinian biology" (Huxley 1966), "the workshop of evolution" (Nash 2008) and "Darwin's fabled, Equator-hugging evolutionary petri dish" (Bellows 2006; see also the more detailed history of the laboratory metaphor in Chapter 4 of Cairns, 2011). The tensions implicit in this rhetorical evocation of science are exemplified by Boersma et al. (2005), who called attention to the threat posed by incompatible goals in the Galápagos (discussed above) and concluded with a clarion call:

"How can scientists help? ...[W]e can call on establishments such as the IUCN [International Union for Conservation of Nature] to help the Ecuadorian government carry out science-based conservation... Our responsibility as scientists is to alert institutions, governments and the public to the de-evolution of the Galápagos islands. The international science community must garner strong global political support for the natural wonders of the Galápagos. Only then will this laboratory of evolution have a chance to persist for another 100 years."

I argue above, as Lawton did before me, that conservation, however science-*supported*, cannot be science-*based* – its foundations lie outside of science, in cultural or, in fact, metaphysical assumptions that we dearly need to make evident. That this is the case is supported by the tangled set of claims that follow in Boersma et al. "De-evolution" would still be evolution; after all there's no requirement that evolution be directional and indeed no criterion – *other than our own* – which could properly disclose that direction in the first place. Evolution often favors simplification, whether of organisms or ecosystems, sometimes even to a monstrous extent (as discussed in Chapter 1). Darwin's observation that introduced species regularly displaced endemic organisms on islands, in the process sometimes reducing diversity[14] and truncating food webs, was one of the facts on which he built his theory,

---

[14] Of course a "reduction" in diversity is necessarily in the eye of the beholder; nothing less than the distinction between same and other is at stake. Seldom do scientists take cognizance of this great difficulty – how similar do you have to be to be the same, how different do you have to be to be different? Instead it gets subsumed by the seeming clarity of taxonomic categories. Even then, however, the problem re-emerges: for example, an organism introduced to a new place may lead to a reduction in the number of recognized species there while at the same time an increase in the number of recognized genera, as Darwin noticed and endeavored to explain (The Origin, Chapter IV, pp. 114 – 5).

arguing as it did against immutability and local perfection.[15] Similarly, how are we to reconcile the "natural wonders of the Galápagos" with the islands being a "laboratory of evolution"? Do we not already see here a repetition of man's dual relationship with nature, as a part of it and apart from it? When is a laboratory natural? Laboratories were developed in distinction from nature, as places where human beings could better control variables otherwise profligate in the natural world. That laboratories remain natural nonetheless can certainly be conceded (owned) – but only if the enforced human isolation from nature is simultaneously conceded (disowned). Otherwise, for the Galápagos to be a laboratory, we would need to protect them from the complicating incursion of nature; but for them to be natural, we would need to protect them from the complicating incursion of ourselves. We have only rediscovered the same metaphysical bind of corruption and redemption from which we started.

Absent an intended distinction between corruption and redemption, if the islands really matter for the insights into evolution which they can provide, then scientifically why would we privilege the insights that pertain to "native"

---

[15] He actually understood it as a step *towards* perfection, the conquest of the less perfect by the more perfect. Thus, the yielding of endemic island fauna and flora to species introduced by humans was evidence, for Darwin, that "man has unintentionally stocked [oceanic islands] from various sources far more fully and perfectly than has nature" (Chapter XII, p. 390). Likewise, he considered the biological productions of Great Britain "higher" than those of New Zealand precisely because of the former's success in becoming naturalized in New Zealand and "exterminat[ing] many of the natives," which circumstance he could not imagine were New Zealand productions to be loosed in Great Britain (Chapter X, p. 337).

organisms more than to "introduced" ones? "I am surprised that the introduced species' significance as a scientific resource has been recognized in so few cases" wrote Kramer (1984) of the Galápagos,[16] before outlining specific evolutionary or ecological questions that introduced species could help answer. Again consider the case of the maligned black rat. Clark (1980) speculated that "[t]he invasion of *Rattus rattus*, which unfortunately ended the first [taxon] cycle by eliminating most endemic rodents, may mark the beginning of a new cycle of diversification and speciation" – precisely as Darwin would have predicted. Indeed, earlier, in their examination of *R.*

---

[16] Such a charge could not be levied against Darwin, who routinely referenced introduced or naturalized organisms to prove central claims of his theory, including (to name just a few instances) the tendency of living things to increase geometrically (Chapter III, pp. 64-5), the still-imperfect adaptation of living things to each other and to their physical conditions (Chapter IV, pp. 82-3 and again Chapter XII p. 390), the relationship between increased area and greater speciation and the related susceptibility specifically of islands to biological invasion (Chapter IV, p. 106), the absence of a known limit to species richness (Chapter IV, p. 110), the impossibility that "the same species can be produced at two separate points" (Chapter XI, p. 352), the immense importance and geographical scope of migration (Chapter XI, p. 377), the dynamics and outcome of migrations and competition during the Glacial period (Chapter XI, p. 380) and "the principle, that the greatest amount of life can be supported by great diversification of structure" (Chapter IV, pp. 114-5). For Darwin, introduced organisms do not threaten evolution or its "laboratory," as they do for, e.g., Boersma et al. Rather, they are the "experiments," as it were, which provide crucial evidence *for* evolution. On this score, it's also noteworthy that a few pages from the end of his book Darwin opined that "[a] new variety raised by man will be a far more important and interesting subject for study than one more species added to the infinitude of already recorded species" (Chapter XIV, p. 486). Do naturalized rats count as new varieties raised by man?

rattus on eight Galápagos islands, Patton et al. (1975) were able to discriminate three distinct populations, based on analyses of allozymes, nonmetrical variants and mensural characteristics. They were further able to link those populations to probable periods of settlement and to propose whence the rats arrived. (The black rat population on San Cristóbal likely arrived in the early 19th century, they suggested, probably from the west coast of South America – incidentally, the probable source of the somewhat earlier colonists, *Oryzomys*.) Pergams and Ashley (2001) re-examined a subset of Patton et al.'s data (those from Baltra and Santa Cruz) in the context of microevolution of island rodents occurring over a relatively short time (defined by them as the span of a human life).[17] They documented significant morphologic changes on the Galápagos – noses became longer and wider while skulls became more shallow – which they argued were likely due to microevolution rather than phenotypic plasticity. More surprising still was that very similar changes were observed in other rodent populations (house mice and deer mice) on distant islands – a Welsh island, a Scottish island and three California Channel islands. Overall, extremely high rates of microevolution were estimated for a majority of mensural traits.[18]

---

[17] Here again we observe the "incursion" of the human into what purports to be an objective measure: for why is the scale of a human life appropriate to an understanding of microevolution? But to what is micro- or macro-evolution in reference, if not us or the standards we select?

[18] Would Darwin be surprised? "Hence it is the most flourishing, or, as they may be called, the dominant species, – those which range widely over the world, are the most diffused in their own country, and are the most numerous in individuals, – which oftenest produce well-marked varieties, or, as I consider them, incipient species" (Chapter II, pp. 53-4). What is he

For those with a disinterested interest in evolution, a historically-constrained group of rats apparently exhibiting relatively high rates of microevolution oddly in step with changes observed elsewhere might seem like a bonanza, as recognized of *R. rattus* populations in other contexts – e.g., the "unique ecological resource" represented by *R. rattus* on the Shiant Islands, Hebrides (Key et al. 1998) or in the study of founder effects on the Guadeloupe islands (Abdelkrim et al. 2005; as also advocated in general terms for the Galápagos by Kramer 1984). Indeed, Pergams and Ashley, citing a paper from 1904, note that "[s]tudies of island rodents were among the first recognized examples of microevolution." But they could have cited Darwin's fumbling intimations of that already in The Voyage of the Beagle (p. 379):

"At James Island [= Santiago], there is a rat sufficiently distinct from the common kind to have been named and described by Mr. Waterhouse; but as it belongs to the old-world division of the family, and as this island has been frequented by ships for the last hundred and fifty years, I can hardly doubt that this rat is merely a variety produced by the new and peculiar climate, food, and soil, to which it has been subjected."[19]

---

describing if not rats? See also his description of rats and mice, which live "free under the cold climate of Faroe in the north and of the Falklands in the south, and on many islands in the torrid zones" (Chapter V, p. 141). That they accomplished this by availing themselves of human-supplied transportation does not seem to have disqualified their achievement in Darwin's eyes.

[19] Darwin later discounted the role of purely physical conditions in driving this variation, emphasizing instead the greater role of the mutual relations

The recent rediscovery of an endemic Galápagos rat, *Nesoryzomys swarthi* (Dowler et al. 2000), known before only from five 1906 specimens and considered likely to be extinct even in the paper first describing it (Orr 1938), not only exonerates the black rat, for now, of its extinction (long attributed to it)[20] but also provides an opportunity for further ecological and evolutionary studies, e.g., of the ≥ 180-year coexistence of *N. swarthi* with *R. rattus* (not to mention *M. musculus*) on Santiago (Harris and Macdonald 2007). (Indeed, Patton et al. [1975] argue that the Santiago black rats came with the buccaneers in the late 1600's, so co-existence may

---

between organic beings – an insight which, in The Origin, he attributed to his Galápagos observations, as discussed in Chapter 1.

[20] It is important to observe that, despite the repeated invocation of science as justification for the extermination/restoration campaigns in the Galápagos, the trial and condemnation of "aliens" frequently proceeds on little more than circumstantial evidence and far-from-rigorous observations and inferences. In the case of rats, this is even admitted, though the admission in no way affects the conclusions drawn. Thus, in describing Project Pinzón's ambitions to kill 180 million rats, Nicholls (2013) reported, "Galápagos experts have little doubt that the rodents have devastated native wildlife, *even though the creatures' effects have not been studied systematically.*" (Italics added.) Why do the experts have so little doubt despite the admitted absence of systematic study? Should they not be chastened already for having blamed black rats for an extinction which, it turns out, hadn't happened? Why, in this legally protected nature preserve where visitors are everywhere greeted by signs reminding them that, under the auspices of the Republic of Ecuador, "the indigenous wildlife of these islands is strictly protected by law," is *habeas corpus* denied rats? Is it because, though they've been there hundreds of years – exactly as long as the human legislators – they're still not indigenous? But why would science care about such a distinction, unless it has yielded to that confusion long accompanying it, between human and natural law?

extend beyond 3 centuries.) And yet, rather than being excited by such prospects in this "living laboratory of evolution," most of the discussion of the black rat in the Galápagos has focused on how best to exterminate it (for a recent summary of which, see Table 1 in Harper and Carrion 2011 and Nicholls 2013).

"One sign of success" of the Galápagos restoration program, according to a CDRS scientist interviewed by Kaiser (2001), was the cooperation of villagers on Santa Cruz in the elimination of "every rock dove they found in one town... except for a few owned by one holdout family." Rock doves have the misfortune of being suspected invasives – and "invasives... are considered the number one threat to the extraordinary diversity of birds and reptiles that inspired Darwin's theory of natural selection" (Kaiser 2001). Any serious student of The Origin knows, however, that the development of Darwin's theory actually depended far more on domestic pigeons (rock doves) – he devoted an entire section of Chapter 1 to them and returned to them numerous times throughout the book as well as in his subsequent works – than on, say, the iconic (and mythologized) Galápagos finches, which he didn't even label properly. ("Unfortunately most of the [Galápagos] specimens of the finch tribe were mingled together," he lamented in The Voyage of the Beagle, Chapter 17). Indeed, Darwin's most detailed discussion of Galápagos birds in The Origin (in Chapter XII) ironically first focused on his astonishment that they should so decisively "[betray] their original birthplace" (p. 399). "Why," he wondered (p. 398), "should the species which are supposed to have been created in the Galápagos Archipelago, and nowhere else, bear so plain a stamp of affinity to those created in

America?" His answer? They're "colonists... liable to modification" (p. 399). But now in this "living laboratory of evolution," and ostensibly in defense of it, colonists are instead liable to extermination.

IV. The Humanistic Case: Cultural and Economic Benefits

Wee, sleekit, cow'rin, tim'rous beastie,
O, what a panic's in thy breastie!
Thou need na start awa sae hasty
Wi bickering brattle!
I wad be laith to rin an' chase thee,
Wi' murdering pattle.
I'm truly sorry man's dominion
Has broken Nature's social union,
An' justifies that ill opinion
Which makes thee startle
At me, thy poor, earth born companion
An' fellow mortal!

(from "Tae a Moose" [To a Mouse], Robert Burns 1785)[21]

So far I have shown that the ecological, evolutionary and conservationist arguments for Pleistocene rewilding and

[21] The poem concludes, "But Mousie, thou are no thy-lane, / In proving foresight may be vain: / The best laid schemes o' Mice an' Men, / Gang aft agley, / An' lea'e us nought but grief an' pain, / For promis'd joy! / Still, thou art blest, compar'd wi' me! / The present only toucheth thee: / But Och! I backward cast my e'e, / On prospects drear! / An' forward, tho' I canna see, / I guess an' fear!"

Galápagos restoration all fail to provide a true foundation. The privileges they extend to specific ecosystems, ecosystem processes, evolutionary potentials, species or species interactions cannot actually be justified on the scientific grounds alleged for them; they cannot be justified by a supposed knowledge. That supposed knowledge instead always takes for granted an unexamined humanism. Hence the justification for those valorizations points beyond the fields in which they are made, and it is there that their true ground must be sought. Presumably that ground can be identified, albeit loosely, as cultural; at least "cultural and economic benefits" are the last of the three legs on which the argument for rewilding would stand. So what exactly are those benefits? How do they support "restoration" projects?

Donlan et al. (2006) presented three at-times intertwined claims, one economic, one aesthetic and one ethical, as follows:

(1) The economic claim was that rewilding would increase the "value of both private and public parks and reserves." They pointed to the economic benefits of wolf reintroduction in Yellowstone, which have been appraised at \$6–\$9 million per year as compared to an estimated annual cost one tenth of that. Much of the projected economic benefit was due to increased tourism, as is especially clear from their Table 1, where tourism was the most common and, in half of the cases, the exclusive economic benefit that they anticipated. (The next most common was hunting, which they restricted to elephants and lions.) Of all of the proposed megafaunal introductions, only that of camels was not listed as resulting in an economic

benefit due to tourism – a surprising omission, since they themselves noted the existence already of camel tours in Texas and even gave the website (www.texascamelcorps.com). As of this writing, overnight trips through the Big Bend area are $750 per person, 3 person minimum.

(2) The aesthetic claim began with our "fascination with charismatic megafauna," for which they cited as evidence "ancient rock art, cars and sports teams named after large mammals, and conservation programs" as well as the 1.5 million annual visitors to the San Diego Zoo's Wild Animal Park (now more sexily known as the "Safari Park"), which is more than visit all but 12 of the 59 National Parks in the United States. With respect to the Yellowstone reintroduction of wolves, they concluded, "[e]ven the possibility of catching a glimpse of a wolf in Yellowstone National Park improves and contributes to the anticipation, authenticity, and enjoyment of a tourist's wild experience. A public understanding of ecological and evolutionary history, inspired by tangible aesthetic experiences with megafauna, would strengthen overall support for the conservation of biodiversity and wilderness."

(3) The ethical claim pertained to human responsibility for the late Pleistocene extinction and for the ongoing plight of large vertebrates, which we ought therefore to redress, and made an appeal as well "for the sake of future human generations and the earth's biota." The only citation to which they referred in support of their ethical assertions was Ehrlich's 2001 article, "Intervening in evolution: Ethics and action." It is notable that

Ehrlich framed his argument wholly in terms of human benefits: "The ethical questions about intervention in the evolutionary process are very similar to questions about the closely related issue of the preservation of biodiversity. Most of us believe that people in the future should be able to obtain from biodiversity a wide variety of aesthetic pleasures, ecosystem goods, and, especially, ecosystem services... Our acceleration of the rate of extinctions and modification of the evolutionary process immediately raises an ethical issue long recognized by economists and others as that of 'intergenerational equity.' The basic question is: At what rate, if any, is it moral for the current generation to discount the future?"

The economic and aesthetic claims are sufficiently entangled that I shall treat them together, beginning with this observation: as argued above, the threat to the endemic organisms of the Galápagos, as in many other places, is not divorceable from the tourist attention that the islands not only gather but seek. If human presence is what knocked nature off its proper course to begin with, it is strange that supposedly knocking nature back onto that course should be justified in part by the opportunities it will enable for human presence.

But what kind of human presence exactly is this? What are tourists? Why would tourism be the privileged form of human participation in nature? Besides the collateral effects of tourism (e.g., the inadvertent introduction of organisms, the occasional oil spill, or the development of supportive infrastructure), besides as well its grossly disproportionate first world audience, tourism carries a more pernicious consequence. So long as human beings understand themselves as apart from nature, as somehow not directly entailed in nature and accordingly beyond nature's purview, it is hard to see how a deep sense of responsibility will develop for the natural consequences, the consequences in nature, of potentially *any* human action at *any* time and in *any* place. But tourism risks perpetuating precisely this illusion of human separability from nature. The tourist, far from being enmeshed at all times in a natural world that affects and is affected by him, instead visits "nature" at his leisure, as his finances permit and as a spectator (at the San Diego zoo, safely ensconced in an aerial tram) – and then retreats from it when the "real" world beckons. Thus Bellows (2006) can describe the Galápagos as "where the family of man meets nature" and Kristof (2005) can argue that "rewilding might entice Americans away from their televisions and connect them with nature." But the global environmental changes that threaten innumerable organisms including, in a real and stark way, human beings, are unlikely to be addressed in any meaningful sense if we think our encounter with nature happens somewhere else than where we are[22] or only in the presence of charismatic megafauna rather than among the ever-less-common bees pollinating the flower that crept through the crack in the sidewalk. In short, they are unlikely to be meaningfully addressed if we accept not only the separability of "nature" and the "human," but also as self-evident that we know what we mean by these terms in the first place. Under such circumstances, tourism risks being an indulgence that absolves us from the consequences of our

[22] Even if we are in front of a television.

actions; it risks being a prophylactic encounter with only a vision of nature.

And it is precisely only a vision, as unwittingly confessed by Simberloff et al. (2011) in their defense of conservation success: "[T]he goal is maintenance of a *semblance* of the original community, with all the species *playing* their *roles*" (italics added). So too Bellows (2006), who, with comparable wit, claimed that "the Galápagos exude a pull that's part Peter Pan, part Jurassic Park" and "just may be the world's greatest family amusement park" while incongruously adding shortly afterwards, "the real thing abounds in the Galápagos." "[W]e don't need a fake in a land that preserves so many originals," he chirped, of course not wondering at all what he meant by "originals" and probably unaware of the proposals to replace extinct tortoises on Pinta Island with close relatives (e.g., Hamann 1993). Which only raises the question: what is real and what is fake here? Have we truly been enticed away from our television sets to connect with nature, when that experience of nature is forced into our aesthetic preconception, partially shaped by television, of what it ought to be?[23] Voilà, a world deliberately enriched with charismatic megafauna – but not with amphibians or invertebrates; a world where the possibility of glimpsing a wolf gives the Yellowstone tourist's

---

23 Again I think of Hawthorne's "The Birthmark," particularly the accomplishment in Aylmer's laboratory of a simulacrum more beautiful than the world: "The scenery and the figures of actual life were perfectly represented, but with that bewitching, yet indescribable difference which always makes a picture, an image, or a shadow so much more attractive than the original." Notably (and necessarily) these "airy figures" are described as "absolutely bodiless ideas" – which announces the latent Cartesianism which I shall shortly address.

"wild experience" greater "authenticity,"[24] whatever that means – but where dandelions are weeds, rats are tramps, and pigeons are aliens or junk; and a world where Project Isabela receives major funding from the tourism company Lindblad Expeditions, to whom in turn the project dedicates its thematic atlas (Lavoie et al. 2007). The genius of reality television was to realize that the script was superfluous, having already been internalized. Is the genius of restoration biology any different?

Consider how the Galápagos Islands have been portrayed historically. In the first recorded human visit, exactly 300 years before Darwin's, the Bishop of Panama described the islands as "like a slag, worthless" and complained that they looked as though "God had showered stones," an observation all the more poignant because of the Spaniards' then desperate need for fresh water.[25] David Porter, the US Naval officer who visited the islands in the early 19th century, described them in 1815 as having "the most dreary, desolate and inhospitable appearance imaginable" (cited by Cairns 2011, p. 66). For generations, the islands were used as penal colonies infamous for their misery. Darwin saw some of those prisoners when he visited in 1835. His own impressions of the islands were hardly more favorable than those of his forebears. In his diary he noted the comparison of the islands "to what we might imagine the cultivated parts of the Infernal regions to be" and continued in The Voyage of Beagle (pp. 374 – 5):

---

[24] See the end of Chapter 4 for a more probing critique of the invocation of authenticity.

[25] http://www.galapagos.to/TEXTS/BERLANGA.HTM

"Nothing could be less inviting than the first appearance. A broken field of black basaltic lava, thrown into the most rugged waves, and crossed by great fissures, is every where covered by stunted, sunburnt brushwood, which shows little signs of life. The dry and parched surface, being heated by the noonday sun, gave to the air a close and sultry feeling, like that from a stove: we fancied even that the bushes smelt unpleasantly. Although I diligently tried to collect as many plants as possible, I succeeded in getting very few; and such wretched-looking little weeds would have better become an arctic than an equatorial flora. The brushwood appears, from a short distance, as leafless as our trees during winter; and it was some time before I discovered that not only almost every plant was now in full leaf, but that the greater number were in flower."

Likewise Herman Melville, who visited the islands in 1841 and described them in "The Encantadas" (1856):

"Take five-and-twenty heaps of cinders dumped here and there in an outside city lot, imagine some of them magnified into mountains, and the vacant lot the sea, and you will have a fit idea of the general aspect of the Encantadas, or Enchanted Isles [= the Galápagos]. A group rather of extinct volcanoes than of isles, looking much as the world at large might after a penal conflagration. It is to be doubted whether any spot on earth can, in desolateness, furnish a parallel to this group… But the special curse, as one may call it, of the Encantadas, that which exalts them in desolation above Idumea and the Pole, is that to them change never comes; neither the change of seasons nor of sorrows... Another feature in these isles is their emphatic uninhabitableness... Man and wolf alike disown them. Little but reptile life is here found... In no world but a fallen one could such lands exist... However wavering their place may seem by reason of the currents, they themselves, at least to one upon the shore, appear invariably the same: fixed, cast, glued into the very body of cadaverous death." ("Sketch First", pp. 99 – 100.)

In the 19th century, the islands were populated with "wretched-looking little weeds" and cursed by changelessness. By the next century, however, "weeds" referred to introduced species and threatened a changelessness which had become desired. Now the islands were a veritable, or at least literal, Eden. In fact, a farm named "Eden" was established on Floreana in 1929 (Cairns 2011, p. 86). That onomastic inspiration has since been copycatted by oceanographers (who christened a vent field on the Galápagos rift "Garden of Eden"; Corliss et al. 1979), biologists (who perceived "one last chance" to bring the Galápagos "back to Eden" at a CDF and WWF conference on Galápagos biodiversity; CDF and WWF 2002) and, of all people, Richard Dawkins (who referred to the islands as "a geological and biological Eden"; as quoted by Hennessy and McCleary 2011). Most recently, a 2013 feature-length documentary film is entitled "The Galápagos Affair: Satan Came to Eden." Even the islet formerly known as Isla Eden, so named in 1846 to honor Captain Henry Eden, was in 1989 officially rechristened Isla Edén by the Ecuadorian government to make its paradisiacal association clear – though that association had long been erroneously taken for granted, as by William Beebe, who in his 1924 book Galápagos: World's End opined that the islet was especially "well named."

Edenic rhetoric is not limited to the Galápagos, however, and is commonly encountered in conservation projects more generally, as noted by Lawton (1997):

"[T]here is a fundamental, and frequently unrecognized dilemma at the heart of conservation efforts, which the heat of battle tends to obscure, rather than to illuminate. What do we wish to conserve, and why? A simple answer is that we want to reduce, or eliminate the impacts of human beings over at least part of the globe, to conserve remnants of what is 'natural.' Implicit, and sometimes explicit in these wishes is the assumption that there exists some pristine, Garden-of-Eden-like state for all ecosystems, from which they have been disturbed by human actions. But there never was a Garden of Eden that we should strive to save or recreate."

In the Galápagos the rhetoric has gone a step further than Lawton described. For these "antediluvian" islands would be, not a potential Eden, but an actual one. Through a curious alliance between tourism and science – of which a dramatic example is the symbiosis between the scientific extermination campaign Project Isabela and the ecotourism company Lindblad Expeditions – the 20th century thus saw the Galápagos emerge as a "scientific, personal and spiritual paradise" (James 2010; see also Hennessy and McCleary 2011 and Cairns 2011, Chapter 4, for detailed accounts). Only a century before they had been likened to hell. In the intervening time the islands miraculously underwent a veritable theological (in fact

Christian) transformation, from infernal and fallen, to paradisiacal. But this paradise is no less precariously kept than its Biblical precursor; it too is at risk of being lost. "Paradise in Peril" read the title of Benchley's 1999 *National Geographic* article about the Galápagos. "There is a great danger that the archipelago will become another 'lost paradise,'" warned de Groot (1983). Pezzati et al. (1998), nettled and stung with pismires, and wary of the parsimonious emmet, echoed de Groot in their paper entitled "Ants of Floreana: Lost Paradise?" And, assessing Ecuador's 1998 Special Law for the Galápagos, Hoyman and McCall (2013) couldn't help but ask, "Is there trouble in Paradise?"[26]

The aesthetic commitments to restorative conservation projects thus appear to go beyond the exemplarity of rock art, zoo visitations and sports teams named after large mammals. In particular, one can't help but notice that there are *stakes* implicit to these commitments, a possibility that the examples put forward by Donlan et al. (2006) might fail to suggest. Would either Pleistocene rewilding or Galápagos extermination projects be conceivable outside of a Judeo-Christian framework in which human beings are both masters of the natural world and, especially because of their moral deficits, alienated from it? Yet, notwithstanding their common foundation, the hellish and the paradisiacal conceptions of the

[26] One should note that rewilding projects (as also de-extinction campaigns; see the Parodos) also fall into a Christian rhetoric, particularly of resurrection; e.g., Donlan and Martin (2004), the last sentence of which reads almost poetically, "We can resurrect and along the way help save the Proboscidea." For similar rhetoric, see, e.g., Seddon et al. 2014 and Stokstad 2015.

Galápagos fall short of being quite equivalent. On the one hand, this transformation calls attention to the remarkable fact that our perception of the Galápagos has a history. That this is so should challenge the reigning self-evidence of our current valorization of them. On the other hand, we are still left with the mystery of how and why hell became a fragile Eden.

But perhaps that metamorphosis is not as mysterious as it seems. In the 19th century, the Galápagos were a place of toil where the visiting human being confronted his embodiment in the stark form of need. "Cursed is the ground for thy sake; in sorrow shalt thou eat of it all the days of thy life." Beneath the unrelenting Equatorial sun, this jagged, volcanic, reptilian land barren of fresh water and inhabited by prisoners reduced human beings to their materiality: "for dust thou art and unto dust thou shalt return." But the 20th century saw a technologically mediated apotheosis of the islands which paralleled that of man. Human need could, for a time, be subordinated to human desire; improved technology could make the Galápagos more easily habitable, or at least represent it as such; and Darwin and the islands together could be idolized if not fetishized as the source not only of the theory of evolution, but of a progressivist interpretation of evolution that saw human beings as its pinnacle, accelerating towards environmental independence and ultimately fated to control the course of evolution itself. A champion of this interpretation was also a champion of the Galápagos: Julian Huxley, Fellow of the Royal Society, first Director-General of UNESCO, founding member of the WWF, and first honorary president of the Galápagos' own Charles Darwin Foundation. "As an outcome of Darwin's work, we have begun to grasp the central and all-important idea that man is the latest and highest type produced by the evolutionary process, and that his destiny is to guide its future course on this planet," he wrote in the 1966 *Proceedings of the Symposia of the Galápagos International Scientific Project*. Human beings, released from their materiality, their existence as *res extensa*, instead exalted *res cogitans* and their likeness to the divine;[27] and in near lockstep, infernal islands became celestial. "The mind is its own place, and in itself / Can make a Heaven of Hell, a Hell of Heaven," observed Satan in Milton's *Paradise Lost*.

Further contributing to science's role in romanticizing the Galápagos were collecting and other scientific expeditions, especially as they became ever more publicized in the early 20th century (James 2010; Cairns 2011, p. 77). Even before Darwin's, expeditions had sought specimens for the Linnaean project of ordering the biological world, thereby to reveal the Book of Nature and the latent divine plan. In the process, they frequently piggy-backed on, and contributed to, political and socioeconomic programs of 18th and 19th century imperialism (e.g., the involvement of Joseph Banks, Daniel Solander, and Johann and Georg Forster in Captain Cook's voyages; or Darwin's own involvement aboard the surveying ship *HMS Beagle*). These collecting expeditions additionally became associated with institutional and often personal prestige. Thus, to name only a few, the Hopkins-Stanford Expedition of 1898/1899, the enormous California Academy of Sciences Expedition of 1905/1906 and the Crane-Pacific Expedition of the Field Museum of Natural History in 1929, plus William

[27] These Cartesian terms will be discussed shortly.

Beebe's two Galápagos expeditions in 1923 (with the motto "It's all for science!") and 1925. Many of these were routinely covered in the media: e.g., the California Academy of Sciences expedition was followed by the *San Francisco Chronicle* (which had also covered their prior 1903 expedition to offshore Mexican islands; James 2010) and even Louis Agassiz's earlier trip led to an 1873 article by his wife in *The Atlantic Monthly.* Beebe's visits, however, established a new standard for media coverage, carrying along with them reporters, photographers and filmmakers, and culminating in the books Galápagos: World's End (an instant bestseller) and The Arcturus Adventure, whose popular receptions in turn contributed to several attempted Galápagos colonizations (Cairns 2011, p. 77). After World War II, yet more recognizable figures entered the action: the 1953 Norwegian Archaeological Expedition was led by Thor Heyerdahl, while in 1954, the Walt Disney Galápagos Expedition was launched, followed a few years later by the release of the Galápagos-focused and Oscar-nominated documentary "Islands of the Sea." Even UNESCO's 1957 reconnaissance mission, which led shortly afterwards to the establishment of the CDF and the affiliated CDRS, was accompanied by a *Life* magazine illustrator and photographer (Bowman 1960) – consistent with the CDF and CDRS's charge to help preserve the Galápagos in part because that would bolster their tourist appeal.[28] In the process, these expeditions and their media popularizations helped to transform hell into

[28] The fifth of the five stated goals of the 1957 reconnaissance mission was "To obtain adequate photographic documentation of the islands and their biota for publicity purposes." *Life* magazine also provided substantial financial support for the undertaking. See p. 8 in Bowman (1960).

paradise by allowing a large, second-hand audience to have a disembodied experience of the Galápagos sanitized of all but its adventure and visual wonder.[29]

Thus, science and the glorified pursuit of knowledge abetted the romanticization of the Galápagos and thereby contributed, and continues to contribute, to their currently perceived fragility – as will be the case so long as the Galápagos are exalted as Edenic by both scientists and tourist agencies alike. Tourists and scientists will come, and with them further human settlement, plus the maligned "pests and weeds." But scientific projects may have contributed even more directly to the island's defaunation, through the toll exacted by scientific collecting itself. In a controversial call to document rare species in other, non-lethal ways than through specimen collection, Minteer et al. (2014) drew attention to how collecting may have hastened species decline or extinction, citing as examples the great auk, Mexico's elf owl and several plants. Their charge was disputed angrily by Rocha and 122 co-authors (2014), and on different, very fatalistic terms, by Krell and Wheeler (2014). Yet, in the case of the Galápagos, the charge may carry considerable truth.[30] The most infamous

---

[29] That this experience of the Galápagos appears as a vision is not nothing. As discussed in the next chapter, western epistemology has long privileged sight, which carries with it an illusion of non-involvement (i.e., the illusion of the passive spectator) not paralleled, for example, in the sense of touch or taste. Perhaps sight also lends itself particularly to an imagined disembodiment if only because of the disappearance of the bridge of our nose in our binocular vision.

[30] Indeed this is already acknowledged in Bowman's 1960 report of UNESCO's reconnaissance mission: "Galápagos animals are in great demand by research museums and zoological gardens throughout the world. It is

example is the California Academy of Sciences expedition, which, on the eschatological premise that the Galápagos were rapidly approaching their doom, gathered over 76,000 specimens (James 2010). Among organisms known only from that expedition is the lone representative of a disputed subspecies of tortoise from Fernandina Island, *Geochelone nigra phantastica* Van Denburg 1907. The fate of the Pinzón tortoises is more extreme and yet almost comical: expeditions in 1897, 1898, 1900 and 1901 all declared theirs to be the last specimen; when the California Academy of Sciences expedition arrived on Pinzón in 1905 and found to its surprise 86 tortoises, including over 60 females, still on the island, they duly killed and collected them all (Thornton 1971, p. 137).[31] A similarly grotesque spectacle attended the Academy's expedition in 1903 to the Revillagigedo Islands and collection of the now-extinct San Benito House Finch (James 2010).

---

within the realm of possibility that over zealous collecting by museums and zoos could actually threaten with extinction such animals as flamingos, penguins, flightless cormorants, albatross, etc..." (p. 38).

[31] Let me add one last irony. The massive campaign to kill 180 million rats on Pinzón has been explicitly justified by the threat they pose to this surprisingly resilient (if oft-miscounted) tortoise. Thus Nicholls (2013): "The rodents feed on the eggs and young of seabirds, land birds and reptiles, and have brought several species – including the rare Pinzón giant tortoise (*Chelonoidis duncanensis*) – to the brink of extinction." Simply to blame rats and not humans – and particularly scientists – for the Pinzón tortoise's current plight is either exasperatingly blind or an illuminating expression of a confusion which regularly reigns between rats and men, at least since Hamlet struck behind the arras. For are not rats here treated as metonyms of men? Is not their eradication, no matter its purported scientific justifications, a form of sacrifice, both as atonement and in renunciation of an aspect of ourselves which we would but cannot renounce?

Thus, just as there is a history of how the Galápagos have been perceived and imagined, from hellish to paradisiacal, so too is there a history of how scientists have understood their responsibilities and roles there. They have typically been neither insightful nor critical about that understanding, instead merely being carried along by the ideology of their times, as suggested by one last example: the scientific history of introduced organisms. In the 18th and 19th centuries, explorers under the auspices of the Royal Society and supported by Enlightenment theory, deliberately introduced domesticated plants and animals, including pigs and goats, to 'savage' islands, thereby supposedly helping to civilize their inhabitants.[32] By the early 20th century, in part as a result of

[32] This endeavor found its theoretical justification in Adam Ferguson's 1769 "Essay on the History of Civilized Man," which identified four stages in the progression of man from savage to civilized: dependence on hunting, which gave way to pastoralism, followed by the rise of agriculture and eventually of commerce. Indeed, no less a figure than Benjamin Franklin, FRS, argued in these terms: "[A] commercial nation particularly should wish for general civilization of mankind, since trade is always carried on to a much greater extent with people who have the arts and conveniences of life, than it can be with native savages..." To speed up the advance of civilization, Franklin therefore advocated for "a voyage of subscription, to carry the conveniences of life, as fowls, hogs, goats, cattle, corn, iron, etc..." to these still-savage places (Franklin, as quoted in Fry 1979, p. 48). 18th century navigators dispensing herd animals and seeds and laying the grounds for commercial exchange could thus understand themselves as catalyzing progressive human development. For example, Captain Cook would write, apropos his third voyage, that "I could not help flattering myself with the pleasing idea that some future navigators may, from the same station, behold the meadows stocked with cattle brought to these islands by the ships of England; and that the completion of this single benevolent purpose, independently of all other considerations, would sufficiently mark to posterity that our voyages had not

those introductions, an apocalyptic and self-fulfilling notion that "time was running out" for native plants and animals justified their slaughter on an impressive scale for preservation in museum and academic collections (James, 2010). And by the early 21st century, the same apocalyptic notion now justifies, again as preservative, both the introduction of organisms (under some circumstances) and the eradication of introduced organisms (under other circumstances). Some will attribute these shifting goals to the progress of knowledge – although the entire argument of this chapter disputes that. Rather, these shifting goals appear as the expression of changing, never explicit and basically acritical aesthetic and economic commitments, to which the professed pursuit of knowledge has in fact been liege.

Those commitments are also claimed to be ethical. Pleistocene rewilding would be an ethical atonement. On the one hand, an obvious and basic question is overlooked: is atonement possible? Or would redress merely be the

---

been useless to the general interests of humanity" (Cook and Barrow, p. 28). He added, "the satisfaction that I felt in having been so fortunate as to fulfill his Majesty's humane design, in sending such valuable animals [including pigs and goats] to supply the wants of two worthy nations, sufficiently recompensed me for the many anxious hours I had passed..." (Cook and Barrow, p. 29). The humanity of these designs was often rather tortuous (or even torturous): in Eimeo, when natives stole two goats intended instead as benevolent gifts, Cook undertook a small-scale military operation in order to retrieve the animals and, in the words of surgeon's mate David Samwell, "punish[ed] [the natives] severely for their treachery in order to put a stop to any further robberies" (as quoted by Hough, p. 304). The robberies threatened the principles of commercial exchange, on the one hand, and on the other, the high-minded, disinterested benevolence favoring the progressive development of humanity.

imposition of another simulacrum ("proxies"), one that immediately effaces the ancient but still unthought violence ostensibly motivating it?[33] On the other hand, why do human beings, alone of all the myriad organisms which, in the history of life, have contributed to the extinction of their neighbors, bear moral responsibility? Because we are conscious of our actions? Are we? Is it ever this clear and distinct? It is as impossible for me to believe that our Pleistocene ancestors were in any sense aware that their individual actions were leading to the defaunation of an entire continent, as it is for me to hold to account ancestors of the microorganism *Methanosarcina* for their acquisition of methane-producing genes and possible contribution thereby to the Permian-Triassic extinction (Rothman et al. 2014). It is also far from clear to me that we are conscious of our actions even now, as my discussion of the changing roles of science above suggests. When a tourist inadvertently carries a seed or an insect in the cuff of his shirt, has he simply acted consciously? When a scientist with seeming deliberation manipulates an ecosystem and thereby unleashes a chain of unforeseen consequences, has she simply acted consciously? When groups of people unite to try to realize in the world an ideological vision which they have no inkling is ideological, have they simply acted consciously? And when 7 billion people collectively release greenhouse gases that dramatically change the planetary climate, fish fisheries to the brink of collapse, and inaugurate a planetary mass extinction likely in the not-far-distant future to

[33] It may also efface the ongoing violence against less charismatic organisms.

include them (Schramski et al. 2015), have they simply acted consciously?

These questions resonate in part because they call attention to the very framing of the ethical problem: at stake are the ethics of "intervening in evolution," as Ehrlich (2001) described it. Already we should pause and wonder: how are human beings and evolution being imagined such that the one *intervenes* in the other? The perspective is that of an evolutionary tourist safely ensconced in his tram. It is a Newtonian perspective, a god's eye view overlooking the vistas of absolute space and of absolute time. But the remove of a tram is illusory; they are neither perfectly safe nor without effects in the world which they would inertly oversee. Newton's absolute space and absolute time are also illusory, as Einstein demonstrated in 1905.[34] And a position of non-involvement, from which evolution's unfolding can be assessed and, if need be, directed, is also an illusion. The ethical question is not really that of intervening in evolution – in which we are wholly and inextricably embroiled; rather it is, how do we justify what we privilege, the organisms we seek to assist as opposed to those we seek to hinder, moreover in a context of ever-imperfect knowledge and ever-incomplete consciousness?

Ehrlich's answer to this question explicitly privileges the human: we owe future human generations "aesthetic pleasures, ecosystem goods and, especially, ecosystem services." What if anything we owe the non-human – whether future, present or past – is unaddressed. Donlan et al.'s (2006) answer on the

[34] This will be discussed further.

surface might seem more satisfying: they justify rewilding "for the sake of future human generations and the earth's biota." But the seeming greater generality of their answer for its inclusion of "the earth's biota" is an empty generality: the problem is precisely that some members of that biota will necessarily be privileged at the cost of others. How to decide who wins and who loses? Nor is it clear that the interests of future human generations and "of the earth's biota" – if this is even a meaningful concept – will align. Certainly the interests of future human generations, assuming that those interests are not so heterogeneous as to preclude convergence in the first place, will conflict with the human-assessed interests of at least some of the future earth's biota; how could they not? Finally, the temporal framing is also a problem; as discussed earlier, the temporal horizon of an evolutionary strategy, if such ever really exists, or of a conservation plan, is far from self-evident. But is it any more evident how far into the future our present ethical responsibilities extend? And if we owe something to a future that doesn't and may never exist, do we owe anything to the past that no longer exists?

To see some of the stakes of these ethical questions we can once again turn our attention to the Galápagos eradication projects. In the absence of those extermination campaigns, it is likely that many endemic organisms will decline and eventually go extinct, as seems to have been the case for, e.g., several of the endemic rats on the Galápagos once their Old World cousins arrived (Brosset 1963). Even if we don't privilege endemic organisms, it is clear that the success of some groups, like goats, will necessarily come at the expense of other groups. Voilà Darwin's struggle for existence and all the important sequelae that come of it. This is what it means to be finite, in a finite world, with finite resources. Perhaps this is what the glass walls of the tram would protect us from. How comfortable, then, can we be in saying the goats must die so the tortoises may live? As much as we might like such a judgment to be the output of a scientific and preferably mechanized calculation, how can it be? It is a value judgment; and science, laden as it is with values, only ever inherits them. The brief scientific history sketched above also militates against the simple-minded solution of pretending that science can resolve this for us; science, after all, has been the celebrated handmaiden of many an undertaking we now revile. Thus, that the project, whether of Galápagos eradication or of North American rewilding, is "science-driven" is both misleading and ethically irrelevant – or, if anything, calls attention to the amoral facility with which science can contribute to any endeavor.[35]

---

[35] Very similar issues are currently playing out in the campaign to "eliminate all invasive vertebrate predators" from New Zealand (an island of 268000 square kilometers and including major human population centers) by 2050, as (uncritically) discussed by Owens (2017). Spearheading the campaign is a scientist, James Russell, whose own career, ironically if not allegorically, began with the deliberate introduction of rats to islands in order to study their establishment. Among the targeted invasive vertebrate predators are again rats (despite one species of rat having been on the island since the 12th century), possums (deliberately introduced for the fur trade but now known as a vector for a disease threatening island cattle), and stoats (deliberately introduced in the 19th century as a biocontrol of rabbits and rats). Though today's problems are clearly and undeniably the consequences of our deliberate past interventions, no doubt justified then by our seeming knowledge, it passes without comment that of course this time we really know what we're doing. Even while the motivations are characteristically

It is hard not to hear the echo of that historic facility in the purification projects now underway in the Galápagos – perhaps especially after one of them (Project Isabela) explicitly claimed it "did not require ethics approval" (Carrion et al. 2011). What science has contributed to these projects is not in fact their justification, despite failed attempts, but rather their means of achievement, meticulously documented and with particular attention to economic cost. When faced with the question, "*Should* we kill every single introduced goat, pig,

---

confused: explicitly justifying this project are the benefits it would have for agriculture (which largely entails similarly introduced plant and animal species, including the cattle indirectly threatened by the possum) and for tourism, which recently eclipsed agriculture as New Zealand's primary source of revenue. Government officials, Owens writes, "worry about the hit to the country's reputation as a destination for unspoilt beauty" that invasive vertebrates pose – even while efforts are simultaneously being directed toward protecting against the collateral loss of introduced game animals (pigs and deer) as part of mass poisoning campaigns. Happily, as the minister of conservation pointed out, "we're good at killing things." She added that, to get over the hump of killing every last straggler, "we'll rely on scientific breakthroughs" – breakthroughs such as gene drive technology for the deliberate genetic engineering of a species into extinction. Surely in all of this no unforeseen consequences will result; no new "pest" will rise up to replace its extinguished competitor, no untargeted organisms will inadvertently be exterminated, no population of desired species will crash due to an indirect reliance it developed on an invasive predator, and no genetically engineered suicide modules will be mobilized beyond their intended population. Such questions, as always, need not overly concern us; nor need we be chilled by Russell's enthusiastic comment that "we're in a relatively unique position in New Zealand, where people are really, really willing to kill for conservation. It's kind of a national pastime." After all, he personally is "proud to look back and see how far we've come in just ten years" – the ten years which have passed since his own deliberate introduction of rats to islands in the interest of science.

donkey, rat or ant on a given island?" science can properly offer, as argued above, no answer. But science has much to say when instead the question is, "*How* can we kill every single introduced goat, pig, donkey, rat or ant on a given island?" Here the team of scientists working with Josh Donlan on the Galápagos has been at once inventive, cost-effective and methodical, in several cases detailing the extermination of particular unwanted organismal groups to both the very last animal and the very last dollar. The scientific contribution is thus one of instrumentality, a role that would be foisted onto ethics as well: during Project Isabela, Carrion et al. noted, hunters were trained "in all facets of hunting and field skills, including dog handling, ethics, GPS, radios, rifles, first aid, and telemetry." Ethics apparently is a tool or a skill, like GPS or a radio or science itself. But understood in this way the ethical no less than the scientific can only fail in providing direction. Both are means; and means do not justify ends.

They do, however, obscure the implications and the ideological foundations of ends. Hence, passing unremarked and almost certainly unnoticed by Carrion et al. (2011) is the chilling resonance of their "restoration" project with the political purification programs of the early and mid-20th century. For those programs too justified enormous violence against specific groups on the basis of both historical and natural ideals, in particular on the basis of an imagined past purity sullied by uninvited, undesired and lesser races ever threatening a degrading homogenization.[36] Further, those

[36] It is sobering here to recall ecologist William Laurance's already-cited (p. 209) critique specifically regarding the Galápagos that "What we're doing is

programs explicitly invoked Darwinian doctrine and supposed scientific advances in support of their actions, into the efficient implementation of which science likewise was entrained. And, like the CDF's first honorary president, the eugenicist Julian Huxley, they understood man to be in the last stages of achieving mastery not only over nature but over evolution itself.

Lest I be accused of throwing around such accusations lightly, let me call attention to the very language in which the Galápagos campaigns are described. It is the language of war.[37] "Galápagos takes aim at alien invaders," read Kaiser's (2001) piece, in which goats are further described as "marauding troops." Nicholls (2013) described Project Pinzón – which he called a code-name – as "a military-style plan-of-action to kill

homogenizing the world's biota; setting the world on a geological epoch: the Homogocene."

[37] It is also the language of hate. Consider the following (part of which I have already quoted), from Nicholls' (2013) description of Project Pinzón: "Galápagos experts have little doubt that the rodents have devastated native wildlife, even though the creatures' effects have not been studied systematically. 'I just hate the immigrant killers because I could see what they were doing,' says Felipe Cruz, a lifelong conservationist who grew up on Floreana, one of four inhabited islands in the archipelago." Cruz, of course, is a frequent collaborator of Donlan and co-author with him on many of the vertebrate extermination projects undertaken in the islands. One wonders a number of things: Why are the rats immigrants while Cruz himself is a native? Are all immigrants killers, or just some? Does Cruz hate the immigrant killers because he can see what they are doing, or does he see what they are doing because he hates them? (Isn't this the sort of pitfall systematic study is supposed to avert?) Finally, most seriously: Is ecological nativism the correlate of political nationalism? With what confidence and with what forgetfulness would we keep the ecological and the political realms separate?

invasive rodents." Anchorman Dan Rather likewise warned (quite confusingly) in 2008 that "[a]n important battle is being waged between Nature and Man here in the Galápagos" (as quoted by Hennessy and McLeary 2011).[38] Guo (2006) portrayed scientists on a "search-and-destroy mission" in a Galápagos "under siege." War is also the rhetoric used by the scientists themselves. Kramer had already declared the islands "under siege" in 1983. *Nature Conservancy* specialist Juan Carlos Gonzalez claimed that "this is a very expensive but totally necessary war" (Associated Press 2012). "A great battle has been won here," declared the aptly-named Victor Carrion, first author or co-author of the cited donkey, pig and goat extermination papers, in an interview with Guo (2006). And when scientists deployed radio-collared "Judas" goats to find the few remaining goat survivors on northern Isabela island, thence to kill them, a subset so-deployed were sterilized females put into a prolonged state of estrus and named *Mata Haris* "after the seductive and lethal World War I double agent" (Lavoie et al. 2007). It is scarcely surprising, then, when Jerry Guo (2006), in his coarsely titled *Science* article "The Galápagos Islands kiss their goat problem good-bye," concluded his piece by unwittingly identifying the restoration project with Nazi Germany and the atrocious 20th century history whose evocation has long been just beneath the surface: "To avoid ceding hard-won *breathing room* for native species, the park and research station plan to set up a $15 million fund for ongoing eradication efforts. In the meantime, they are stepping up efforts against invasive plants and gearing

38 The quotation is confusing in part because, in this war between Nature and Man, it's not clear which side Man is on.

up for the [planned] cat-and-rat *blitzkrieg.* Unless these and other unwelcome visitors go the way of the goats, warns Carrion, 'the worst may be yet to come.'" (Italics added.)

Were we to imagine the worst, including whether or not it is yet to come, would it include the individual perspectives of 201,825 beings whom we happen to call goats, each one killed ultimately for an idea? The idea, first, that each is a goat; second, that goats are introduced, invasive, alien; third, however nebulous, that through their deaths a purity, even a paradise, is regained. Would it include the perspectives of the Judas goats who, after the trauma of capture, sterilization and, in some cases, terminated pregnancy, "escape" to the wild and to the company of other goats, only to endure, like caprine Typhoid Marys, the yet worse trauma of having every companion they encounter killed before their eyes? "3,439 feral goats were shot while associated with Judas goats," Carrion et al. (2011) reported with cold precision, unaware that their very terminology made a Christian pageant of their undertaking – made each goat betrayed by Judas a sacrificial lamb.

Would it include the perspectives of individual pigs, individual donkeys, individual black rats and, yes, even individual little fire ants? Would it include the perspectives of each endangered tortoise, each endangered Galápagos rail, each endangered endemic rat and the endangered *Scalesia* too? For the worst, were anyone ever even to conceive of imagining it, begins from this aporia, this resourcelessness in the face of inevitable suffering and inevitable complicity in suffering.[39]

---

[39] And it begins from a confrontation with absolute particularity that at best I have only gestured at, by emphasizing (problematically) the individual (as if I truly know what that is). One major part of the problem, which cannot

That resourcelessness, suffering and complicity are all effaced in these projects, however – as are, in a curious way, the killed animals themselves. Seldom is it even explicit that the animals are being killed in the first place.[40] Table 1 in Carrion et al. (2011) neatly tallies the number of goats, not killed, but "removed" from each island, to the grand sum of (I will repeat it again) 201,825, corresponding to a no-less-neatly tallied cost of $11,958,282. But the goats, like the pigs and the donkeys and the rats and presumably even the ants, were not removed. Their carcasses were left to rot and fester where they fell. Also absent from the account: the terror of the hunted animals (who "quickly become educated and wary, often hiding in bushes, caves or lava tunnels," Carrion et al. state); their death throes, whether quick or prolonged; their bloated bodies; the stench of their partially scavenged corpses. The

---

be divorced from the pursuit of knowledge as such, is our inescapable reliance on language and hence on generalization. (This is the problem which drove Aristotle to invent secondary substance in the Categories, primary substance being literally unspeakable; see Chapter 2, p. 168.) After a while we forget that "goat," "pig" or "rat" are our categories, rather than fundamental truths of the world; and we certainly forget that the being before us may not understand itself in the terms by which we understand it. What is this forgetting but the subtle exaltation of our ideas toward universality, the highest claim of knowledge? – even as we would leave behind an awkward, fragile being that is not even and certainly not simply, e.g., a goat. Or a human.

40 This is true as well of the earlier collecting expeditions, such as the California Academy's. This became clear to me during the question period after a public lecture, when an audience member's question revealed the assumption that collected tortoises were merely taken into protective custody, rather than killed.

animals are instead treated as if disembodied – as if what was killed was an idea all along.[41]

V. The Post-humanistic Case: Memento * * * * *

"...*I have seemed to see, slowly emerging from those imagined solitudes, and heavily crawling along the floor, the ghost of a gigantic tortoise, with 'Memento * * * * *' burning in live letters upon his back.*" Melville, "The Encantadas," Sketch First, pp. 102 – 3.

Where are we left?

If I have succeeded, I have shown that the restoration projects for North America and the Galápagos are not actually justified on the ecological, evolutionary or ethical grounds asserted for them, and that the economic argument, among other problems, privileges a peculiar if not pernicious form of

---

[41] Again the figure of Hawthorne's Aylmer comes to mind – he whose scientific library was "full or romance and poetry," the examples of which apparently are Albertus Magnus, Cornelius Agrippa, Paracelsus and Roger Bacon. And his own laboratory notebook: "The book, in truth, was both the history and emblem of his ardent, ambitious, imaginative, yet practical and laborious life. He handled physical details as if there were nothing beyond them; yet spiritualized them all, and redeemed himself from materialism by his strong and eager aspiration toward the infinite." As with rewilding and the purification of the Galápagos, Aylmer's project is redemptive – redemptive of materiality, finitude, mortality, if nothing else. In the case of his wife, he would ultimately cure her of her very embodiment: in his laboratory, he treats her as a material thing, the symbol of that materiality being, in fact, her only blemish. *But he does so in order to spiritualize her.* Thereby is she, like those introduced blemishes in the Galápagos, both scapegoat and sacrificial lamb.

presence – tourism – which will only perpetuate the very processes and attitudes that it deplores. The aesthetic argument is more complicated. On the one hand, the particular ways in which these projects, places and beings have been aestheticized contributes to the very desirability which is simultaneously understood as threatening. On the other, the aesthetic appropriations have changed over time: predators that in the Pleistocene were probably objects of fear and awe are now objects of pity, wolves are no longer devilish, and the fallen or infernal Galápagos are revealed instead as paradisiacal or celestial. Such transformations suggest something more basal underlying them, and so the aesthetic argument, like the others, points beyond itself to a foundation that so far has not been articulated.

That foundation is metaphysical. One approach to the metaphysics at stake in these projects has already been well-discussed by Hans Jonas (1966) in sections III and IV of "Is God a Mathematician?" There he describes how the Judeo-Christian alienation of god from the world, in which the latter is created and not creator, material and not divine, provided a template for the stark but essential division between mind and nature that reached its most important expression in the Cartesian *res cogitans* and *res extensa*. The resultant extreme form of dualism supplied, he argued, "the metaphysical charter for a purely mechanistic and quantitative picture of the natural world" (p. 72) – in short, for the scientific enterprise as we know it. But also enabled with Descartes was the unacknowledged and sly abstraction of human beings from their materiality and their contexts even – and indeed precisely – for investigations of the physical world. Henceforth, the merely material world could nonetheless be known from an immaterial but still ostensibly human perspective. For Descartes, it was the perspective of the *res cogitans* and its uniquely human affinity with the divine; it was a god's-eye view become humanly accessible.

Yet it was precisely this god's-eye view that the revolutions of early 20th century thought – notably in physics, anthropology and phenomenological philosophy – all challenged, always by instead emphasizing human embodiment and the inescapability of context. Thus, Einstein's 1905 theory of relativity is predicated on the spatial and temporal situatedness of all human activity, including the supposedly inert activity of observation. What started as a minor asymmetry in the explanation of electromagnetic phenomena culminated in the overthrow of the seemingly secure and separate realm of mechanics – all because that realm had failed to take into account the observer as embodied and, therefore, necessarily both somewhere and somewhen. It was only by acknowledging that situated embodiment, that materiality, that Einstein succeeded in describing how two observers could correctly explain the same phenomenon in fundamentally different ways. Since he could find no criterion by which to privilege one observer's account over the other's, however, Einstein relinquished the very possibility of that privilege – and hence of absolute space and absolute time. Similarly, in 1927 Heisenberg rationalized his Uncertainty Principle as also resulting from how human ways of knowing unavoidably affect what is known.[42] Virtually at the same time

[42] The Uncertainty Principle is no longer rationalized in this way.

(1929), it was recognized and formalized by Bronisław Malinowski that anthropologists were no less constrained than physicists to be participants in, as well as observers of, the activities which they recorded. Phenomenological approaches in philosophy also called attention to the importance and inescapability of human situatedness, for example in Heidegger's Being and Time (1927), with its emphasis on facticity and *Dasein* – not merely being (*sein*) but being-there (*da*) and being-in-the-world.

But this situatedness is conspicuously absent in the conservation projects that I've described. As their very rhetoric suggests, they would return to or safeguard an Edenic refugium where the materiality of human beings can either be compensated for or ignored. Theirs is a Newtonian vision, a dream of a presence that is not anywhere.[43] Thus is reconciled the overt paradox of human beings as both corrupters and redeemers: man as *res extensa* would corrupt; but man, now as *res cogitans*, would redeem.[44] Operating as a sort of behind-the-

43 I associate this with Newton because his assumption of absolute space and absolute time requires a further but more covert assumption of a disembodied and atemporal viewer. It was in effect this covert assumption which Einstein challenged.

44 In Hawthorne's "The Birthmark," this radical dualism is represented by the laboratory assistant Aminadab, whom Aylmer refers to as "thou human machine... thou man of clay," and by Aylmer himself as thinking substance: "'Ah, clod! ah, earthly mass!' cried Aylmer [to Aminadab, if perhaps to others as well], laughing in a sort of frenzy, 'you have served me well! Matter and spirit – earth and heaven – have both done their part in this." The narrator concurs: "[Aminadab] seemed to represent man's physical nature; while Aylmer's slender figure, and pale, intellectual face, were no less apt a type of the spiritual element." Spiritualized man of course soon is confused with god, and indeed Aylmer emerges as a god-like object of worship for his

scenes ghost-in-the-machine, human beings would thereby direct the ecological order without being a part of it.[45] It is a macabre but comical vision, one which, on the Galápagos, entails killing hundreds of thousands of organisms,[46] while a like number of human beings wander the islands, carefully "put[ting] everything, every pebble, back where it was" (Galápagos National Park tour guide Klaus Fielsch, as quoted by Bellows [2006]) if accidentally disturbed because nature, after all, shouldn't include them. But this god's eye view,

---

wife and for himself ("'Ah, wait for this one success,' rejoined he, 'then worship me if you will. I shall deem myself hardly unworthy of it.'"), while, were she not marked with the sign of her mortality, he in turn "would fain have worshipped" her as well. Yet, however much Hawthorne invites the separation of matter and spirit, *res extensa* and *res cogitans*, through the personifications of Aminadab and Aylmer, he also disrupts that possibility absolutely, not only in the person and fate of Georgiana, but notably through Aylmer's laboratory journal. This writing reveals that the man of science and his Igor-like assistant are not opposite poles after all: for it "was the sad confession and continual exemplification of the shortcomings of the composite man, the spirit burdened with clay and working in matter, and of the despair that assails the higher nature at finding itself so miserably thwarted by the earthly part."

[45] No greater critique of this view of theater – one in which the separation of the performance from its institutors and audience is supposedly clear and distinct – can be found than in Hamlet, when the Danish prince, long suspicious that his globe is a theater (Act II, scene ii), strikes through the arras and kills Polonius, who is at once rat, actor, stage director and audience. So is the conceit abolished that the theatrical and real worlds are separated absolutely; and the directors, actors and audiences who fail to rise from their seats in alarm have failed to understand what even Claudius could not overlook. The most astonishing and disregarded accomplishment of theater is thus our belief in the absolute border represented by what is, after all, only a curtain.

[46] In fact, millions upon millions.

which in a benighted world is strangely equivalent to the view of a tourist ensconced in an aerial tram, is achieved only creatively, only imaginatively – through "proxies," as Pleistocene rewilding advocates. It is achieved only as "a semblance… with all the species playing their roles," to quote again Simberloff et al. (2011). The human ascension is a stage trick, a *deus ex machina*; and all the world's a stage. Bucking the overt trend of the early 20th century, its roots lie instead in the early 17th century, with the Shakespearian pretender, Francis Bacon. For, Bacon promised, from his method "the benefits of discoveries may extend to the whole race of man… [D]iscoveries are as it were new creations, and imitations of God's work." Those artful imitations, he continued in the 129th aphorism of Book I of The New Organon, will "be great enough to justify the saying that 'man is a god to man.'" Thereby will "the human race recover that right over nature which belongs to it by divine bequest." Bacon's too was avowedly a project of Judeo-Christian simulation and thereby restoration.

In this context, it is hardly surprising that the ascension of the *res cogitans* entails the exaltation of knowledge, albeit often understood in an impoverished way. The dreams of Pleistocene rewilding, Galápagos purification and similar undertakings would reduce those projects simply to tasks of knowledge. Hence the obligatory promises that all will be "science-driven." Hence the ultimately empty recognition that "a full understanding of the relevant natural history, as well as the values of the people with a stake in the outcome, will be essential to any path forward" (Tewksbury and Rogers, 2014). Hence finally the disturbing – because doctrinary or even indoctrinary – invocations of education: to reduce reintroduction risks on the Galápagos will require "a longer term social approach focused on education," Carrion et al. (2011) stated. The "momentous obstacles" associated with reintroducing lions to North America "would have to be overcome with a detailed science-based management and education program," wrote Donlan et al. (2006). And of course ecotourism is routinely, if parenthetically, justified as educational: "There's no more wondrous place on Earth [than the Galápagos]" proclaimed Bellows (2006), "for families to amuse (and educate) themselves." The glorification of knowledge also underlies, and perhaps peaks in, Julian Huxley's progressivist interpretation that evolution is now man's to direct. When Huxley warned, in the same 1966 essay extolling Darwin and the Galápagos cited earlier, that the net effect of medical and social improvements is "almost certainly dysgenic and man's evolutionary fitness is being reduced," it is undoubtedly to knowledge that he would turn to counter this threat – even if, characteristically, it was also knowledge that led to the improvements and their undesired consequences in the first place.

In general, then, whatever risks that are recognized, whether in an ecological or an evolutionary project, are to be skirted with the aid of knowledge; it is simply a question of being fully aware. Peter Kramer, the former director of the CDRS and long-time president of the CDF, argued as much in his provocatively entitled paper "Man and other introduced organisms" (1984):

"[M]an… is the dominant species on the Galápagos. I do not deplore that. There is no reason why he should not dominate and manage all the resources of this earth for his purposes. My impression is that the danger is not caused by man's domination, but by his inability to fully assume his role as a responsible dominant – in spite of his unique ability to act consciously. In a way he continues to act as though there are still ecological constraints, as though Mother Nature is still there to put him in his place for his own good."

On the one hand, Kramer can't help but (disturbingly) to echo Bacon here, who lauds, again in the 129th aphorism, the "endeavor to establish and extend the power and dominion of the human race itself over the universe." For Bacon, this is the third grade of ambition, "more wholesome" and "more noble" than either the vulgar desire of a man to dominate his nation or the higher goal of laboring for one's nation that it may dominate mankind. As Bacon's series makes clear, and as is suggested by Kramer's language of domination as well, this project is also innately political. And like all political projects, it involves the performative.[47] If really an endeavor of knowledge at all, it will accordingly be a knowledge not

[47] By "performative" I am trying, in a highly provisional way, to designate an alternative to a statement that is simply true or false. A promise, for example, can be interpreted as a performative statement because its veracity (or lack thereof) is conditional on subsequent actions. Note, however, that this distinction is tenuous at best. For example, we might think the statement "That animal is a cat" is not performative; either the animal is or is not a cat. The statement is either true or false. In so deciding we quickly forget the problematic natures of the categories "animal" and "cat" and our own involvement in not just discovering but creating them.

merely uncovered but manufactured and performed.[48] On the other hand, what would be the condition of possibility for man to achieve "a *full* understanding" (Tewksbury and Rogers, 2014) or "to *fully* assume his role as a responsible dominant"? Surely man would require complete knowledge, complete aware-ness, complete consciousness. That knowledge would have to include knowledge that is performatively produced; it would also have to include knowledge of its own metaphysical underpinnings. This is an infinite – indeed, transcendent – project requiring an infinite – indeed, transcendent – knower. Yet we are not that. We *do* have a material existence in the world, and there *are* constraints, intellectual as well as ecological. If knowledge is potentially infinite, we nonetheless are finite. Our cup can only runneth over.

But evolutionary theory as articulated by Darwin makes this critique better than I can. As is well known, Darwin developed his theory of natural selection by analogy to the methodical selection practiced by human beings. Hence the greater importance of pigeons than of Galápagos finches in The Origin. Often overlooked in his argument, however, is its seeming reliance on an intermediate term between methodical and natural selection: unconscious selection. "The man who first selected a pigeon with a slightly larger tail, never dreamed what the descendants of that pigeon would become through long-continued, partly unconscious and partly methodical selection," he wrote in the first chapter (p. 39). One point is that man's preferences, including aesthetic ones into which he lacks insight, have material consequences in the world – just

[48] Here we begin to see the inseparable entanglement of the political and the aesthetic.

like those of birds do. But a second, more important point, often wholly overlooked, is that methodical selection is never so methodical that it isn't also unconscious.[49] In consciously privileging one trait, we unconsciously privilege all the traits that come with it. We never dream – or else we only dream – where this will end. Thus to criticize Darwin's theory of natural selection on the grounds that nature can't be an agent gets everything upside down. Implicit to the critique is the concession that man *can* be such an agent. But Darwin's text is more circumspect; not only does it explicitly state that unconscious selection is more important than methodical selection (on p. 34), but it so regularly pairs the two, as in the passage quoted above, that one might wisely question whether ultimately the evanescing term isn't unconscious selection but methodical selection.

Put otherwise: what is suspect in the theory of natural selection is not simply the agency of nature but the agency of man and ultimately the agency of agency. So far was human control of evolution from Darwin's imagination that instead, at the end of Chapter III (p. 78), as I have already noted, he emphasized the necessity of embracing ignorance: "It is good

[49] Indeed, one of Darwin's key insights, a tent-pole of his theory, is "that community of descent is the hidden bond which naturalists have been *unconsciously* seeking" (Chapter XIII, p. 420, italics added). A few pages later (p. 425) he added: "...[M]ay not this same element of descent have been unconsciously used in grouping species under genera, and genera under higher groups, though in these cases the modification has been greater in degree, and it has taken a longer time to complete? I believe it has thus been unconsciously used..." Note how highly methodical investigations, for Darwin, are unproblematically coupled with a lack of awareness – indeed are founded and organized around such a lack.

thus to try in our imagination to give any form some advantage over another. Probably in no single instance should we know what to do, so as to succeed. It will convince us of our ignorance on the mutual relations of all organic beings; a conviction as necessary, as it seems to be difficult to acquire."[50] Surely we are far from convinced of our ignorance on the mutual relations of all organic beings when we contemplate stocking North America with elephants or the Galápagos with Australian ladybugs. But why? Why is this conviction both necessary and difficult to acquire? Its necessity stems from the reluctant admission of our finitude; being finite we cannot encompass the world just as we cannot avoid affecting it. The risk, however, is that, in lieu of this admission, we replace the world with our idea of it. And even this is done in ignorance, without our noticing. Therein appears the difficulty: for that ignorance, that essential inability to be fully aware or fully conscious or even to be, fully, challenges the ultimate priority of the subject. A subject that does not know what it does or why it does it appears instead as an object. And it is toward this odd and contradictory agency that Darwin pointed. By pairing the methodical with the unconscious, he stumbled upon an alternative to Cartesian method and its contraband

[50] By the sixth edition, Darwin had modified this sentence in telling fashion: "It is good thus to try in imagination to give to any one species an advantage over another. Probably in no single instance should we know what to do. This ought to convince us of our ignorance on the mutual relations of all organic beings; a conviction as necessary, as it is difficult to acquire." In the first edition, instead of "ought," Darwin used an indicative future: "It *will* convince us of our ignorance…" Moreover, then the necessary conviction only *seemed* difficult to acquire. Both changes indicate Darwin's increased awareness of how difficult the acquisition of this conviction actually is.

accomplishment of a *cogito* in the first person singular active voice. Darwin's is instead a construction of uncertain agency and person, neither active nor passive but middle.

What happens is always in the middle voice. But the metaphysics subtending Pleistocene rewilding and Galápagos purification is one which privileges the active voice, the subject, the agency of man, and his ability to be fully present (as mind, as *res cogitans*) while obscurely not being present at all (as body, as *res extensa*). It is a metaphysics which makes of the world a simulacrum – a Paradise, an Eden, or equally a Hell – and treats of all the beings in the world as if they were mere ideas. Goats become scapegoats and extinct animals proxies. In this world, no loss is irretrievable, and the Proboscidea eagerly await their resurrection. Perhaps this is the world where the vigorous, the healthy, and the happy survive and multiply – so long as they are native. But it is also a world where the sixth mass extinction and climate change hold an unreal status as facts which we profess to believe without really believing. For to believe them would mean to believe in man's encroaching finitude; it would mean enduring that our ownmost potential for being inescapably includes the potential for not being. It would mean recognizing that when we strike behind the arras, it may not be, and eventually will not be, a rat who dies. But the metaphysics in which we are steeped is predicated on denying all of this. It is the metaphysics of a subject – while we are never fully subjects. It is the metaphysics of presence – while we are always partially absent. It is the metaphysics of the actual infinite. But we are actually finite.[51]

---

[51] Or even, our finitude encroaches us in such a way that we are not even fully actual.

This chapter is striving to recognize a different world, one in which there is loss and suffering and inevitable complicity in suffering, and where the human response to that is not from a transcendent and universalizing perspective of knowledge, presence and infinitude, but rather from a situated position of limitation, entailment and mourning, one in which the dream of rewilding gives way instead to the necessity of bewilderment. When Donlan et al. (2005) asserted that "environmentalists are easily caricatured as purveyors of doom and gloom," they should have hesitated over whether that message wasn't, in fact, in some way both right and important; whether, in fact, our responsibility, in lieu of a violent embrace of an illusion, begins by reconciling ourselves, if that is possible, to the inevitability of absolute loss. Our fates are no different than the goats' or the tortoises'. *Cui bono* must eventually yield to the better though not perfect question *quo vadis*. And yet the question, what are we to do in the Galápagos? – is not thereby evaded. It endures, and we must endure it – an aporia that offers us no absolution from decision. Indeed, the question, what are we to do on the earth? – endures, and we must endure it – and it too is an aporia that offers us no absolution from decision. I have no resources with which to respond to these questions – I am bewildered – but a response must come, is coming and has already come.

In our response may we not fail to read the live letters burning on the back of a gigantic tortoise whose ghost we have seemed to see. They are only partially legible. Remember, they say – but remember what?

Chapter Four, on Science in General
Being Fragments: Acrisis[1] in Modern Science

"Once upon a time, in some out of the way corner of that universe which is dispersed into numberless twinkling solar systems, there was a star upon which clever beasts invented knowing. That was the most arrogant and mendacious minute of "world history," but nevertheless, it was only a minute. After nature had drawn a few breaths, the star cooled and congealed, and the clever beasts had to die."

Nietzsche, "On truth and lies in a nonmoral sense" p. 1

I. Introduction

Too much has likely already been written about Alan Sokal's hoax, his duplicitous publication of a nonsense paper in a non-refereed academic journal that, by its own description, is "interventionist" and publishes "political opinion and cultural analysis" (Robbins and Ross 1996). As a physicist's experiment in Cultural Studies, to paraphrase Sokal's own account of his little adventure (Sokal 1996), it left much to be desired, not least a meaningful argument that the journal *Social Text* was an appropriate proxy for whatever Sokal meant by "Cultural Studies." Either Sokal used the term "experiment" rhetorically, perhaps to privilege his status as a physicist in what would then be a fallacious argument by authority, or it was simply a poor experiment, uncontrolled,

---

[1] How much difference an absent blank space makes! Rather than a crisis, this chapter describes acrisis – a crisis negated by a privative, as, for example, in "acritical."

unrepeated and deploying a metric of undemonstrated relevance to what it purported to measure. One has better things to do than to pay attention to such work.

And yet here I am paying attention to it. Sokal's hoax cast the humanities into crisis; but this is a state in which we frequently find them. Even the age of Shakespeare was, according to his contemporary Richard Burton, a "bastard age of banished muses." Crisis and its etymologic cousin criticism in fact may be good for the humanities, if approached with care and thought, as occasionally they are. It is my contention (with the same caveats) that crisis would also be good for the sciences. Let me go further: the sciences are already in crisis, and an important, even essential, aspect of that crisis is that they have not recognized it, and perhaps cannot.

Here Sokal can serve as exemplar. According to Sokal (in his reply to Robbins and Ross 1996), "a scientific world-view… is based on a commitment to logic and standards of evidence and to the incessant confrontation of theories with reality." Was his own "modest" experiment consistent with that professed worldview? I have already noted that his standard of evidence is at best questionable. How about the incessant confrontation of theories with reality? "There *is* a real world," Sokal continued (the italics are his); "its properties are *not* merely social constructions; facts and evidence *do* matter." Besides an ignorance or intellectual complacency that fails to recognize the deeply legitimate questions that this opening clause can't help but invite (what precisely is meant by "real," "world," or "is," to say nothing of the "there"?), Sokal confused a straw man for what he would undoubtedly accept as a real one; he was tilting at a false world, one he had imagined and confused for the real world. For example, Derrida's work – one of the obvious targets of Sokal, who referred to him several times in his nonsense paper and who began his supposedly more serious confession by announcing that, being a mere physicist, he was "unable to make head or tail of *jouissance* and *différance*" (Sokal 1996) – could never, except through gross ignorance or immense bad faith, be reduced to a mere dismissal of facts and evidence. Sokal didn't do his homework. He didn't bother to verify that his parody was in fact of the world, or at least that the aspect of the world to which it applied was the aspect to which he applied it. (There is of course plenty of nonsense published in the humanities, of which Sokal's hoax is both parody and example.) In short, Sokal was hoist with his own petard and blown at the moon. Illogically convinced by a poor standard of evidence, he was carried away by his own unrecognized rhetoric and resultant confusion of theory for reality.

But I called him an exemplar, and not simply of poor scholarship. I take him to be an exemplar of the crisis in science – a crisis whose symptoms include poor logic, poor standards of evidence and, especially, poor reading. It would hardly surprise Plato that, whether as scientists, humanists or scholars in general, we continue to be liege to our own rhetoric and to a world imagined. Perhaps it would be more surprising to Francis Bacon, he who in many ways inaugurated the endeavor which has become modern science while simultaneously warning, in The Great Instauration (p. 29), that "...all depends on keeping the eye steadily fixed upon the facts of nature and so receiving their images simply as they are. For God forbid that we should give out a dream of our own imagination for a pattern of the world." Yet, Bacon was dreaming here, not only of his explicit project to restore man to his prelapsarian dominion, but of a simplicity of reception in which images appear as they are and merely await their ordering in a table of instances. Sokal continues in this dream. Modern science does too, as I shall try to show. The worldview that science would profess may very well be as Sokal stated – but wasn't his point that a worldview should not be confused with the world?

This chapter will be divided into two parts. In the first I will present what I will provisionally call the symptoms: the various ways in which modern science requires no hoax to parody itself. Such symptoms include the proliferation and dilution of authorship, the concomitant reliance on authority, the "massaging" of data, the failures of peer review, reproducibility and readership, the always anomalous scandal, and the proposed solutions that include further mechanization of scientific inquiry, experimentation and even reading. Each of these symptoms has been individually remarked by other authors, although to my knowledge they have never been collated and analyzed together. Instead, these symptoms are typically confused for the underlying pathology, which in turn is almost always reduced to various social tensions crisscrossing the practice of science in a way at once regrettable and accidental. Institutional reforms, renewed dedication and greater vigilance would then be curative. Yet, as I will argue in the second part of this chapter, among themselves these symptoms betray a remarkable coherence that belies their interpretation as accidental. I will therefore try, with help particularly from Heidegger, to offer a different and comprehensive interpretation, one which both exposes and tries to understand the necessity of these "accidents." To succeed would then not be to infect the sciences with crisis so much as to reveal that they have always already been infected, that this virus has long been a part of them – has perhaps always and necessarily been a part of them. Understood in this way, the pathological metaphor can only break down, lest we mistake as disease something fundamental to what modern science is – and something which, I fear, precludes the possibility of crisis ever appearing as such in science. Such a conclusion will bring us to the brink of a precipice: if crisis cannot appear as such in science, is it truly self-evident that it can appear anywhere else? This question will compel us to relinquish even a certain Heidegger. It will compel us to plunge.

Part One: Automata in raincoats and other degrees of fraud

"But then were I perchance to look out my window and observe men crossing the square, I would ordinarily say that I see the men themselves... But what do I see aside from hats and clothes, which could conceal automata?"

Descartes, Second Meditation

I. Reproducing and Representing

Let's start with the most dramatic problems, the cases of demonstrated scientific fraud. The last three decades have seen several extraordinary examples, including in the late 1990's the

94 papers by cancer researcher Friedhelm Herrmann that an investigative committee deemed "definitely or highly probably manipulated" (Abbott 2000), 59 of which were coauthored by distinguished senior researcher Roland Mertelsmann; the 1999 discovery by Victor Ninov and 14 coauthors of elements 116 and 118, based on fabricated data; in the early 2000's, 28 papers by physicist Jan Hendrik Schön – 20 of which were coauthored by his eminent colleague Bertram Batlogg – retracted as fraudulent from among the world's most prestigious scientific journals (e.g., 9 from *Science*, 7 from *Nature* and 6 from *Physical Review*); shortly thereafter, the faked stem cell research of Hwang Woo-suk, which also involved egg donations collected if not coerced from junior scientists including graduate students (Cyranoski 2004) and again ensnared a prominent journal (*Science*) and a prominent researcher, Gerald Schatten, who remarkably tried to renounce his coauthorship retroactively (Elias 2005); 21 papers revealed in 2009 to be based on data fabricated by anesthesiologist Scott Reuben, many of which were sufficiently influential that they contributed to attempted pain management of real patients (Borrell 2009); similarly, in 2012, the 172 papers containing false data written by another anesthesiologist, Yoshitaka Fujii, but happily little read (Akst 2012); the 150 papers by psychologist Diederik Stapel that, in 2011, came under scrutiny for "made up or manipulated data" (Vogel 2011), including an eventually retracted paper in *Science*; and the recent retraction of a physics paper again from *Science* due to unspecified "inappropriate data handling" by lead author Rabindra Mahato (Mahato et al. 2015). This is far from an

exhaustive list; these are only among the most high-profile examples.

Indeed, research misconduct appears to be on the rise. Using data from the National Institute of Health's Office of Research Integrity, Check (2002) documented a near-doubling of reported misconduct, between 1998 and 2001. Steen (2011) described something similar: although he determined that, according to the formal retraction notices, most retractions (73.5%) between 2000 and 2010 were due to error rather than fraud (26.6%), he also found that both the number of retractions and the incidence of fraud rose sharply in that time period. A more comprehensive study by Fang, Steen and Casadevall (2012) partially reversed Steen's earlier conclusion, which had relied on what now emerged as "incomplete, uninformative or misleading" retraction announcements, and instead reported that two-thirds of over 2000 retracted papers published in the biomedical or life science fields were withdrawn due to misconduct, principally fraud (43% of the total retractions). Comparable to the findings of both Check (2002) and Steen (2011), they also documented a ten-fold increase in the percentage of papers retracted due to fraud since 1975, with a marked rise in the frequency beginning in the early 90's. As disturbing as these accounts are, the responses they elicit are often minimizing (Primack 2013, Gunn 2014) or perversely self-congratulatory, with the revelations being taken as proof of science's self-correcting nature: e.g., "the process by which [Schön] was found out shows how science – albeit sometimes slowly and clumsily – inevitably ferrets out such fraud" (Smaglik 2002). Such grand claims (note the "inevitably") would be more comforting were there the slightest available evidence to support them – for example, an estimate, undoubtedly difficult to make (for essential reasons to be explicated below) of the amount of fraud that passes currently undetected. In lieu of such evidence, how can confidence like Smaglik's be distinguished from the mere profession of an article of scientific faith?

Two strands of prima facie evidence in fact ought to challenge that confidence, if not faith. The first pertains to the failure to reproduce or replicate the vast majority of scientific claims, and the second to the pervasion of questionable scientific practices which exist somewhere on a spectrum that includes as one end-member fraud. (The continuity of this spectrum, which Gunn [2014], for example, fails to consider, is what will always problematize an estimate of the total amount of fraud, detected or undetected.) I shall deal with these two strands in turn, starting with the first and showing how the second is related to it.

Reproduction or replication – and in this chapter I shall not distinguish them[2] – is commonly advanced as the ultimate guardian protecting science from both fraud and error. That invocation is often cavalier: for example, according to Harvard

---

[2] Not because I think the distinction valueless, but rather because the papers I am citing commonly treat the terms as synonymous and self-evident. In a certain way, this is symptomatic: should we ever inquire rigorously what we mean by "reproducibility," would we not at some point have to confront, if only as allegory, the confusion of terms whereby replication is sometimes taken as equivalent or nearly equivalent to – that is to say, as a replication or reproduction of – reproduction, and sometimes not? Do we not re-encounter problems *akin* to those discussed especially in Chapters 1 (in the equivocation of self-replication) and 2 (in the "proper function" of the reproductive system)?

immunologist and cancer researcher Fred Alt, "it is impressive that the majority of what is published is reproducible and accurate" (as quoted by Adam and Knight 2002). That would indeed be impressive – but is it true? Start by considering the enormous number of scientific papers that are published per year. In fact, it turns out, for reasons that a moment's reflection will elucidate, this is a very difficult number to estimate, and depends, for example, on how one defines "scientific paper." Thus, Larsen and von Ins (2010) compared the National Science Board's estimate of 710,000 science and engineering articles published in 2005 to another study's estimate of roughly twice that number (1,350,000) only a year later. Reich's 2013 article in *Nature* tallied > 1.1 million articles published worldwide in 1997, and almost double that number (> 2.1 million) in 2012. *The Economist* meanwhile reported in an anonymously authored 2013 piece that 1.4 million scientific articles were being published per year in "scholarly" journals. However uncertain, the number is clearly enormous and appears to be growing (Larsen and von Ins 2010) – probably exponentially (Daruka 2014). Contributing to that growth are the immense and widely-recognized pressures on scientists to publish – future funding and their own promotion may depend on it – which has led some to conclude that "publishing has become an 'inflationary currency'" (Harley 2013). Indeed, with promotion and funding committees, due to lack of both expertise and time, being driven to rely on article metrics rather than content and with career self-interest commonly cited as a major reason for attempted publication (Harley 2013), it is hardly surprising that studies have routinely shown that the fate of the majority of rejected manuscripts is simply to be published elsewhere, often with little or no change (e.g., Ray et al. 2000, Nemery 2001, McDonald et al. 2007). Yet, rather than being alarmed by such findings, or at least asking whether these rejected manuscripts ought to be published elsewhere, several of these studies instead emphasize that their results give rejected authors "reason for hope" (Ray et al 2000; see also McDonald et al. 2007) – thereby again taking for granted publication as the obvious default scenario.

That hope can be fueled by the growing number of journals, which at the turn of the millennium were increasing at an annual clip > 3% (Mabe and Amin 2001) – an estimate which obviously does not include the > 5000 open-access journals started in a mere three year period between October 2010 and October 2013 (Reich 2013). The Directory of Open Access Journals currently lists over 10,000 journals with over 2 million articles. As an example of journal growth, consider the promiscuous spread over the last three decades of the imprimatur of the journal *Nature*: limited to a single journal for over one hundred years after its founding in 1869, it spread to a second journal when, in 1983, *Nature Biotechnology* was launched, followed in the 90's by 5 more journals (*Nature Genetics*, *Nature Structural Biology* [now *Nature Structural and Molecular Biology*], *Nature Medicine*, *Nature Neuroscience*, and *Nature Cell Biology*) and in the 2000's by over 30, including an open-access journal (*Nature Communications*) and additions both in 2016 (*Nature Energy*) and 2017 (*Nature Ecology and Evolution*). Besides journals, new outlets, including conference proceedings, webpages and open archives, are also increasing rapidly (Larsen and von Ins 2010). In the face of this "staggering exponential growth" of published research

(Daruka 2014), some have reached the obvious conclusion: "Scientists today cannot hope to manually track all of the published science relevant to their work. A cancer biologist, for instance, can find more than 2 million relevant papers in the PubMed archive, more than 200 million Web pages with a Google search, and databases holding results from experiments that produce millions of gigabytes of data" (Evans and Rzhetsky 2010). To put the latter number in context, a million gigabytes is roughly an order of magnitude larger than the entire print collection of the Library of Congress. Given these facts, if scientists can't hope to track all of these publications, much less the data on which they are ostensibly based, how can they hope to reproduce the findings?

It might be argued that contributing to the huge numbers of papers being published are replication studies. While technically true, the correction is certainly minor. Historically replication studies have been notoriously difficult to fund and to publish (Collins and Tabak 2014). For example, both the European Research Council and the National Science Foundation have indicated that they would favor "pioneering" work over replication studies when making funding decisions (*The Economist* 2013). As noted by Spotts (2015), "competition is fierce for grant money, which typically doesn't reward redoing someone else's work." Likewise, "studies that replicate (or fail to replicate) others' findings are almost impossible to publish in top scientific journals" (Crocker and Cooper 2011). Replication studies thus do little to promote one's career and may actually hinder it, especially if the results alienate more established (and powerful) scientists.[3] In short, however much the importance of reproducibility is acknowledged in word, in deed it runs afoul of "increased pressures on scientists to advance their research" (Jasny et al. 2011). Nosek et al. (2015) agreed and further implicated "an academic reward system that does not sufficiently incentivize open practices" conducive to reproducibility. Thus there are two problems: first, the enormous number of papers being published and amount of data being produced, which makes reproduction logistically unlikely; and second, an ideology of progress that pervades publication, funding and promotion standards in such a way as to make reproduction, already improbable, also undesirable. Add to that the many realms of scientific inquiry where true replication is fundamentally impossible – whether in field studies that may require immense expertise and can never be sufficiently controlled (Jasny et al. 2011, Ryan 2011), destructive analyses of rare samples, or in large, capital-intensive projects like the Large Hadron Collider or the Sloan Digital Sky Survey, where researchers necessarily lack independent means of generating comparable data (Peng 2011).

In this context, Smaglik's confidence that science "inevitably ferrets out … fraud" and Alt's that "the majority of what's published is reproducible and accurate" look rather credulous. In fact they are: large-scale reproducibility studies

[3] After acknowledging that "incentives to publish positive replications are low and journals can be reluctant to publish negative findings," Baker and Penny (2016) added that "several respondents [to their *Nature*-sponsored survey] who had published a failed replication said that editors and reviewers demanded that they play down comparisons with the original study." Why? Out of a misguided sense of politesse?

have regularly found that the majority of the published scientific works examined cannot be replicated, including in Alt's field of cancer biology. Thus Prinz, Schlange and Asadullah (2011) reported that only 20-25% of the drug target research that they examined was fully reproducible and that reproducibility "did not significantly correlate with journal impact factors, the number of publications on the respective target or the number of independent groups that authored the publication." Likewise, Begley and Ellis (2012), motivated by the "remarkably low" success of translating preclinical cancer research into favorable clinical results, examined 53 "land-mark" studies and succeeded in reproducing the results of only 6 of them. They noted that "some non-reproducible preclinical papers had spawned an entire field, with hundreds of secondary publications that expanded on elements of the original observation, but did not actually seek to confirm or falsify its fundamental basis." The attempt by the Open Science Collaboration (2015) to replicate 100 prominent psychology studies succeeded unambiguously in only 39% of the cases; even then, the effect sizes were typically half of those reported in the original studies. When only the reproduction of data analyses, not data generation, is attempted, the success rate still remains low: Ioannidis et al. (2009), trying to reproduce some of the microarray gene expression analyses found in 18 studies published in *Nature Genetics*, succeeded completely for only 2, partially for 6 and not at all for the remaining 10.

Certainly methodological problems contribute to the difficulty in replication. For example, in a test of liquid chromatography-mass spectrometry-based proteomics, only 7 of 27 labs correctly reported all 20 proteins in a standardized sample and only one listed all of the 1,250 Da tryptic peptides, with the biggest issue having to do not with raw data quality but with different approaches to data analysis (Bell et al. 2009). Similarly, attempted reproduction in the field of cancer biology was often impeded by major disagreements about basic measurements, including proper methods for identification of drug receptors on tumor cells and even what counted as tumor cell growth (Saey 2015). A frequent complaint is also that material and methods sections are inadequate for replication, often because journals limit the space available ("in some journals, it's really a methods tweet," Saey [2015] quoted NIH deputy director Lawrence Tabak as saying) or because scientists have an interest in maintaining a proprietary relationship with their work and therefore are deliberately unforthcoming. (On this latter score, it is noteworthy that neither of the studies by Prinz, Schlange and Asadullah [2011] and Begley and Ellis [2012] is readily reproducible, since these scientists, all but one of whom work for drug companies, withheld fundamental information, including the identity of the research findings which they attempted, successfully or not, to replicate.) Perhaps the most common methodological problem pertains to the misunderstanding and abuse of statistics, whether in early stages of experimental design or in subsequent implementation and analysis (Lambdin 2012, Editors 2014, Nuzzo 2014). Besides pervasive issues of "data-dredging," "p-hacking," "fishing," "significance chasing," etc.[4] – which betray a profound mis-

---

[4] These are common terms for post hoc analyses that in the absence of a hypotheses simply search for statistically significant findings – as if statistical

understanding of typical statistical assumptions – many studies and even fields are characterized by such low statistical power that repeating even a true result may be a shot in the dark. For example, Button et al. (2013) determined that the median power in 48 neuroscience articles, corresponding to 49 meta-analyses and chosen according to preset criteria, was 21% – that is to say, out of 100 true hypotheses, they could expect to detect 21 correctly. Unappreciated by many is the further fact that, as statistical power decreases, not only is the ability to detect a true effect reduced, but so is the probability of a significant result being true (see, e.g., Pashler and Harris 2012). On this score, Gilbert et al. (2016) challenged the findings of the Open Science Collaboration that psychological studies had low reproducibility and argued that the data in fact supported the opposite conclusion, when corrected for error, bias and power; that is, they argued that the Open Science Collaboration's statistical interpretation was mistaken, an argument rejected by the original authors (Anderson et al. 2016).[5]

---

significance were an essential characteristic of a data set, divorceable from context as from any hypothesis or even question.

[5] It would be mistaken to understand this dispute as being reducible ultimately to an argument only about the proper application of statistics. At its core, it confronts, rather, a fundamental issue that is not to be arbitrated at a statistical altar: what counts as replication or reproduction? Anderson et al. (2016) and the Open Science Collaboration (2015) did notice this question, but they resolved it merely by imposing an (inadequate) definition: direct replication is "the attempt to recreate the conditions believed sufficient for obtaining a previously observed finding." In this re-creating they take for granted re-cognition: but, insofar as replication or reproduction inevitably entails difference – as, in a remarkably (but not, of course, entirely) similar idiom, I discussed in the first and second chapters – at issue

Yet even bigger issues must be faced, issues that extend beyond sloppy or erroneous statistics and start to veer toward the questionable scientific practices that I mentioned as the second strand of the prima facie evidence challenging faith in science's self-correcting nature. One problem is the pervasion of bias, defined by Ioannidis (2005) as "the combination of various design, data, analysis and presentation factors that tend to produce research findings that they should not produce." Bias can include selective outcome reporting, repeating or continuing a measurement until it gives the favored result, failure to blind oneself while making a measurement in which one has an interest, data mining, etc. Even with no malice aforethought, human beings may unconsciously steer an experiment toward a desired outcome, if only the desired outcome simply that there be a statistically significant (and hence publishable) result. Under the heading "pathological science," the Nobelist Irving Langmuir described several colorful cases of this phenomenon – an honest scientist acting in good faith and still fooling himself. For example, Langmuir observed one scientist counting hallucinations as scintillations in the service of a preferred hypothesis (Langmuir 1989). Similarly, Lipton (2005), in questioning why we would prefer a predictive hypothesis to one which accommodates evidence, noted that in the latter case, the accommodator can never be

---

will always be the questions, are the new conditions sufficiently alike or relatable to the old? Are the new findings sufficiently alike or relatable to those previously observed? These questions are not ultimately of a quantifiable character – indeed, every quantification itself unavoidably invites, or should invite, these very questions. They are, instead, of a literary character, and a *response* to them may be neither a definitive *answer* nor the final word.

certain that she didn't fudge the data. When Simmons et al. (2011) examined researcher degrees of freedom – i.e., flexibility in choosing dependent variables, sample size and the subsets of experimental conditions to be reported – they determined, for a simulation involving 15,000 samples, that a researcher availing himself of all degrees of freedom would have a false-positive rate of 61% (see also similar arguments in Pashler and Harris 2012). Indeed, on the basis of a relatively simple mathematical model, Ioannidis (2005) concluded that most published research findings are probably false and that "for many current scientific fields, claimed research findings may often be simply accurate measures of the prevailing bias." He considered that this conclusion was likely to extend even to many "classic" and "established" studies. Possibly in support of this, the modeling work of Rzhetsky et al. (2006), on how published scientific statements unduly influence the interpretation of subsequent experiments, explained the extremely high agreement found among published statements as due either to an "optimist's universe" in which published statements reflected a very low experimental error rate (with both false-positive and false-negative error rates < 5%; hence researchers are influenced by prior published statements that are almost always correct) or to a "pessimist's universe" where both error rates exceeded 90% – and, on the basis of further modeling, argued the latter was three times more likely than the former, as Ioannidis on the basis of his simpler modeling work might also predict.

Ironically, a variety of optimism may underlie a major form of bias: the strong preference for positive results, understood as a beneficial and/or significant outcome

(Callaham et al. 1998). In itself this is amusing, given how widespread the ignorance of statistics is (e.g., Nature's editors acknowledged in 2014 that "most researchers do not understand the basis for a term (P-value) many use every day"; see also Pashler and Harris 2012) and arguably, how logically incoherent basic notions like P-values actually are (Lambdin 2012, Goodman 2016). For many, statistics appears to be a modern form of extispicy become the ultimate arbiter of their scientific inferences. Yet, a scientist critical of this usage would nonetheless be foolish not to recognize the extreme difficulty with which "insignificant" or negative results are published – despite the fact that a negative result is typically a more robust finding than a positive one. Part of the problem is that journals "like big, exciting, positive findings," according to Oxford statistician Douglas Altman (as quoted by Couzin-Frankel 2013). Begley and Ellis (2012) argued similarly: "Journal editors, reviewers and grant-review committees often look for a scientific finding that is simple, clear and complete – a 'perfect' story."

If they get that perfect story, however, it may come at a price. When Begley approached a senior scientist about the failure to reproduce one of the latter's findings, the senior scientist, according to Begley, candidly confessed that "We did the experiment a dozen times, got this answer once, and that's the one we decided to publish" (as quoted by Couzin-Frankel 2013). There is thus clearly selective outcome reporting, at two levels – whether data are published at all (studies yielding only negative results commonly are not) and precisely which data are published (within a study, positive results preferentially are). Furthermore, even when negative results are published, they are often spun so as to draw positive conclusions from them, sometimes in tortured ways: Couzin-Frankel (2013) described one case in which scientists compared their treatment group to the placebo group of another study in order to argue for treatment efficacy. One should also observe that the cherry-picking of data implicit to the sort of bias being discussed here has as its counterpart the cherry-picking of literature support, all the easier, and in a certain way even necessary and unavoidable, given the vastness of that literature. Indeed, proposals that all data be published or made publicly available in huge data repositories (Nosek et al. 2015), though well-intended, seem naïve: if there is not enough time, much less inclination, to study and reproduce even the highly select data being published, who will critically peruse what otherwise risks being an information-concealing data dump? These proposals in their naïveté also overlook the profound question of what data, and especially 'raw data,' actually are, when all access to them is unavoidably mediated.

This inevitable mediation comes in view as we realize that these problems of data representation are not limited to the more-or-less unconscious suppression of negative results whether through wishful thinking or as uninteresting or at least unpublishable. After the controversial publication in *Science* of Purdue physicist Rusi Taleyarkhan's claimed demonstration of table-top fusion via sonoluminescence, several of *Science*'s anonymous referees took the extraordinary step of making their negative reviews public. One of them, Seth Putterman of UCLA, testified that he had pointed out how data in the paper disproved Taleyarkhan's claim – and that, in lieu of addressing this objection, the problematic information was simply not included in the final article (Adam and Knight 2002). If Putterman's claim and analysis are both true, this casts not only Taleyarkhan but the editors of *Science* in the rather dubious light under which omission shades into falsification. (Taleyarkhan was eventually found guilty of research misconduct by a Purdue University committee, not however for mishandling data but for having falsely asserted independent data reproduction; Maugh 2008). Indeed, it is unclear, as has been repeatedly recognized, precisely where the line is between 'questionable research practices,' such as mining data for statistically significant findings or selectively publishing data favorable to one's own hypothesis, on the one hand, and on the other, outright fraud. Not surprisingly the former are more customarily admitted to: in a meta-analysis of survey data, Fanelli (2009) determined that on average 2% of scientists self-reported fabricating or falsifying data and/or altering or modifying results, while 9.5% (and as high as one-third) admitted to questionable research practices such as those described above. On average 14% reported observing serious misconduct by others. Similarly, Martinson, Anderson and de Vries (2005), in a survey of > 3000 scientists, found a third willing to admit to having committed in the last three years one or more of ten behaviors likely to be institutionally or federally sanctionable, including falsification (to which 0.3% of respondents admitted), the suppression of data contradicting their own prior research (6%), and "changing the design, methodology or results of a study in response to pressure from a funding source" (15.5%). Additionally, 10% or more admitted to what Martinson et al. considered less problematic behaviors, including "inappropriately assigning authorship credit" (10%), "withholding details of methodology or results in papers or proposals" (10.8%), "using inadequate or inappropriate research designs" (13.5%) or "dropping observations or data points from analyses based on a gut feeling that they were inaccurate" (15.3%). Curiously, Martinson et al. considered the last two behaviors simply forms of "carelessness."

The question is, what underlies these "careless" or "questionable" practices? Typically, as already mentioned, this question is answered at a sociological or psychological level, by invoking, e.g., career incentives to publish. I see no reason to dispute such an explanation while also seeing no reason to find it particularly satisfying. A deeper interpretation begins with the inescapable role of mediation, of which the sociological/psychological explanations ultimately are but examples (i.e., scientists act in a social or psychological context which, at least in part, mediates their judgments). Mediation extends to the representation of all data, including so-called raw data. Even the rawest data are already representations, with all the caveats which that necessarily implies. The very act of measurement is a rhetorical gesture, basically equivalent to synecdoche – in which an indefinitely complex entity or process is represented by one or more of its aspects (e.g., mass, length, etc.). That reduction already entails several mediating judgments – e.g., that the object or process is isolatable in the first place, and that the ways in which the representing aspects are faithful to it are more important than the ways in which they necessarily are not. Those mediating judgments are of course further compounded by whatever means of measurement are employed, as has been recognized since at least

Kepler, who (rightly) obsessed over the legitimacy of the mediations instituted by both telescope and eye (Alpers 1983). The problem is that the object or process never appears simply as itself, not even in the rawest data, an archival record of which is, of course, but a further remove. If this sounds postmodern, it isn't; it has been recognized by empiricists since Locke. Representation is unavoidably at work even before the scientist-author stops to consider how these representations will themselves be represented, first in the data analyses, and then in the public representation of those data analyses in published form.[6]

---

[6] In this light, we can revisit the 2006 paper by Rzhetsky et al. on how published scientific statements influence the interpretation of subsequent work. To cut to the quick: whether they realize it or not, their paper suggests that there is *no such thing* as a truly "independent" observation, scientific or otherwise. All observations occur within, and are mediated by, a context which inescapably influences interpretation. Rzhetsky et al. *describe* this situation even while not deeply understanding it. Thus, in their concluding model for differentiating post hoc the relative likelihoods of the optimists' and pessimists' universes, Rzhetsky et al. warn (p. 4945) that "in our model, experimental data cannot be observed directly by the research community. They have to be inferred from publications." But what can they possibly mean by direct observation, moreover by an entire research community? Is there really that sharp divide between observation and interpretation which Francis Bacon took for granted? Is observation itself not influenced by context? In fact, as Kepler realized in the early 17th century, experimental data can *never* be observed directly, not even by the naked eye. Furthermore, the very act of measurement, I have already noted, is a mediation analogous to a rhetorical trope. The layers of mediation become all the more apparent if we take seriously that "direct" observation must somehow be made available to a research *community*. In short, the assumption that science, under some set of unstated circumstances, is able to achieve independent or direct observation is nothing less than a form of the veiled essentialism discussed

When it comes to that public representation – and, in fact, of course before then – certain, almost universally unexamined aesthetic criteria come into play, in the form of pressures to tell what Begley and Ellis (2012) described more than once as "a 'perfect' story." Couzin-Frankel (2013) quoted an unidentified scientist speaking in similar terms: when asked to explain how he determined which data to report, he answered, "When I take a look at the data, I see what best advances the story." Clearly, then, it would be naïve to think that the story is merely determined by the data, when the representation of data is determined in part by the story! Only more naïve would be to consider this circle avoidable. Instead we should ask what kind of story is it which orients the representation of data and whence come both its assumptions and criteria? What is a perfect story?[7]

Here is where we find common ground between "careless" or "questionable" practices and fraud. At issue in the former are often implicit or concealed representational choices, whether to put a name on an author list, to drop data points, to withhold a methodological detail or even to submit to statistical sorcery. And at issue in the latter are often – the same choices. Jan Hendrik Schön, mentioned earlier (p. 279), was condemned by an Investigation Committee at Lucent Technologies (the parent company of Bell Laboratories, his

---

especially in Chapter 1; it is likewise nothing less than the perspective of the disembodied Cartesian *cogito* discussed in Chapter 3; and it remains latent even in a paper which, at the same time, necessarily challenges it.

[7] Here we confront a tension analogous to that between evolution and perfection described in Chapter 2. The story is simultaneously supposed to advance and be perfect; it is expected at once to be open and closed, to project future progress while being in itself complete.

employer) for behaviors which included 1) failure to keep proper laboratory records, primary data or working devices; 2) data selection and omission in order to match within a few percent expected trends; and 3) the substitution of theoretical curves for experimental ones and the "nontrivial alteration or removal of individual data points or groups of data points" (p. 11 in Beasley et al. 2002, henceforth cited as "Investigation Report 2002"). Yet, among everyday scientists, 1) only 45% of those surveyed expressed willingness to share raw data if asked (down from 60% in 2008; *The Economist* 2013); 2) P-hacking, significance chasing and other "flexible" interpretative strategies culminate in a gross underestimation of the real probability of false-positive findings, universally (and extremely problematically; e.g., Goodman 2016) claimed (for so-called significant results) as only a few percent; and 3) regularly admitted "questionable" (or "careless"!) research practices include the routine omission of data, often in service of a favored theory or hypothesis and presumably justified, perhaps sometimes but not always in bad faith, as "trivial." Schön was no anomaly; he was a caricature of the modern scientist, exaggerating features which were nonetheless already there. And those features are not excisable; they are not accidental blemishes which a good plastic surgeon can remove from the face of science. Rather they are the inescapable companions of all representation, whether under-taken in good faith or bad. Until science confronts and criticizes its own naïve approach to representation it will persist in a ludicrous state of viewing the phenomena I have described, despite the enormity of their scale and prevalence, as unfortunate accidents – yet again; as curiously persistent outliers which nonetheless demand no account; as data the exclusion of which is ultimately trivial.

In this respect, Schön's response to the Investigation Report condemning him is informative. He claimed to be convinced that his experimental results were real "although [he] could not prove this" and explained his conduct in terms of his efforts simply "to communicate the science that described the experimental findings" (Investigation Report 2002, Appendix H). The Investigation Committee in some sense agreed with him: at issue for them were not Schön's claims but rather "the persistent, intentional data manipulation and misrepresentation." This they viewed as a violation of "the fundamental tenet of science that results must be presented honestly." But data can only be manipulated; every representation is also a misrepresentation; and the tenet that the results be presented honestly is the creed of a naïf who has yet to realize that results can only ever be re-presented. This is not to deny the import of the adverb "honestly," of the role of good faith, or of the possibility that some representations are far better than others. But it is definitively to cast aside any pretense to immediacy as a guarantor of scientific truth. That pretense can come only from a willful blindness – the same blindness, in fact, which will sacrifice perpetrators of fraud as "bad apples" while ignoring the continuum of related practices, ultimately unavoidable, to which scientists in large numbers admit and of which their work, implicitly or explicitly, is evidence. Such expressions of faith in science reveal only how science is faith. And they fail to account for the phenomenon unfolding before them. For, when we consider the proliferation of articles, including the oftadmitted practice of double-publishing (i.e., publishing lightly altered versions of the same article in two or more journals, a practice to which 4.7% of scientists confessed in Martinson et al.'s survey), the explosion in the number of journals, the calls for data repositories, the attempts to re-plicate and re-produce – what are all of these if not expressions of the feverish necessity of re-presenting, of, as it were, getting to the bottom of representation by representing once more? And what are they as well if not the pursuit of a perfect story, one which would combine paradoxically both perfection, that is to say completion, and continual, uninterrupted as well as unexamined, progress?[8]

II. Authorship and Authority

A locus in which problems of representation in science take on the proportions of a crisis is authorship. This is already implied in much of what has been written above, particularly regarding the proliferation of journals. That proliferation is only possible if articles can be found to fill them, and clearly they can be – even if it is far less clear if readers can be found for the articles. (The rise of open-access journals whose costs are covered by authors rather than readers may in part be an implicit acknowledgement of this difficulty.) Authorship is also obviously in question when it comes to the public representation of data – the at least ostensible ultimate outcome of a scientific project, and the one in many ways (if only because publications are used as metonyms of productivity and

[8] Again, is this bind not similar to Darwin's in Chapter 2?

thereby as metrics for advancement) motivating some of the dubious practices described above, from massaging data toward significance to double-publishing or publishing rejected work elsewhere.

Schön's case, although not the most dramatic in terms of overall number of publications ("only" 28 have been formally retracted, although several more have been cast into question; in comparison, one of the examples cited earlier involved over 100 publications by a single author), is nonetheless especially instructive. As noted by the Investigation Report, Schön was an unusually prolific writer: in 2001 he wrote a paper on average every 8 days, many of which ended up in the most exclusive scientific journals – journals with rejection rates around 90% – including in that year alone four subsequently retracted papers in each of *Science* and *Nature*. Apparently at the time no one wondered how such prolific productivity, and at such a high level, was humanly possible. That curious incuriosity is especially conspicuous in the case of Schön's coauthors and of the journal editorial offices publishing his work. Of course, in the aftermath of Schön's exposure, both groups have been subject to considerable criticism, as is appropriate, although that finger-pointing characteristically has avoided the most important implications and conclusions. In what follows I will endeavor to bring those to the fore, starting first with the roles and responsibilities of the coauthors.

The Investigation Committee officially and unequivocally exonerated Schön's coauthors: "The Committee finds all coauthors of Hendrik Schön in the work in question completely cleared of scientific misconduct" (Investigation

Report 2002, p. 4). Such exoneration is much more the rule than the exception. Roland Mertelsmann, who was co-author on 59 of Friedhelm Herrmann's 94 suspect papers, was officially cleared; the 14 coauthors of physicist Victor Ninov were officially cleared (Dalton 2002); Hwang Woo-suk's coauthor Gerald Schatten was officially cleared; and the retraction notice of Diederik Stapel's *Science* paper explicitly exculpated his coauthor as "in no way involved in the generation of the data" (Stapel and Lindenberg 2011). Clearly fraudulent papers are, as Rennie and Flanagin (1994) quipped, "orphans," with coauthors routinely not "knowing anything about the fabrications or, indeed, almost anything about the papers." This circumstance appears to trouble very few. Indeed, exonerating coauthors – without interpreting the ultimate implications of such exoneration – is often a higher priority than cleaning up the literature, where disgraced papers frequently persist neither marked nor retracted (Abbott and Schwarz 2002). And yet these exonerations are sometimes remarkably damning. Mertelsmann, for example, while "accept[ing] responsibility for the suspect publications," nonetheless asserted his innocence because he "was only an honorary author, and was therefore unaware of the details of the experiments described" (Abbott 2000). Likewise Gerald Schatten, who has parlayed his involvement in the Hwang Woo-suk affair into street credit as a bioethically-concerned scientist, accepted co-authorship on a paper with Hwang for which "his major contribution… was a suggestion that a professional photographer be engaged" (Pearson 2006). Whatever authorship represented to these scientists – and it clearly *represented* something – it had little to do with responsibility for the conduct of experiments or the claims of a paper.

But that may not be exceptional, if we examine the grounds on which coauthors are frequently exonerated. Schön's case is again paradigmatic. All of his coauthors were, not simply exonerated, but as already noted, "*completely* cleared" (italics added). The committee justified this decision by pointing out that, with minor exceptions, Schön alone had manufactured and operated all of the major devices reported in his papers, that none of his measurements were witnessed by a single coauthor, that no laboratory records, electronic data files or working devices were kept, and that all figures in the various papers were prepared only by Schön. In short, his co-authors were innocent of scientific misconduct because apparently at no point had a single one of them insisted on seeing anything resembling the experimental set-up or initial data upon which the extraordinary scientific claims being made were based. Consider: when Irving Langmuir was skeptical of remarkable results reported at a lecture, he traveled the 150 or so miles between Schenectady and New York to check out the experiment in person – in the 1930's; when Wendell Latimer, likewise in the 1930's, doubted the mysterious Allison effect, he travelled from California to Alabama to observe the demonstrations; and when R. W. Wood was suspicious of René-Prosper Blondlot's newly discovered N-rays, he crossed the Atlantic Ocean to see for himself – in 1904 (all examples from Langmuir 1989). In contrast, Schön's collaborators were never so rigorous, or for that matter curious, as to go from their lab to his – despite not only the considerably greater ease and lower expense of transportation in the new millennium than in the early 1900's but, most damning of all, the remarkable fact that his most frequent collaborators (Christian Kloc, Bertram Batlogg and Zhenan Bao, coauthors respectively on 22, 20 and 4 retracted papers) were at the same institution. If Schön was condemned for showing "reckless disregard for the sanctity of data" (Investigation Report 2002, p. 3), his coauthors' faith in the data surpassed that of Job, who, after all, demanded and obtained a response from God. Even colleagues of enormous experience, such as Bertram Batlogg, according to the Investigation Committee a "distinguished leader" in research of the kind being reported, failed to be, not only skeptical, but sufficiently curious about "perfect numerical symmetry in measured data values" or "data... match[ing] the expected trend to within a few percent" as to demand the opportunity to witness the accomplishment of such outstanding measurements. Batlogg's healthy skepticism was limited instead, in his words, to "probing questions" to which he received "reasonable and satisfactory" answers (Investigation Report 2002, Appendix H). But, probing questions aside, why did he not insist on observing how such incredible data were being generated – if only out of excitement and wonder? Was it because the ideology of progress in science primed him to accept an expected and desired result as, after all, eventually inevitable? Was it relatedly because he was so comfortably conditioned to accept, and thereby confuse, the representations of data and that data themselves always are for the world which they would represent? Rather than moralistically denigrating this comfortable conditioning or faith in progress, would we not be wise instead to inquire into each of their apparent necessities?

In the case of several of Schön's other coauthors, the Investigation Committee did not make such inquiries. It did, however, offer a clear answer to why the coauthors sought no opportunity to observe the initial data: even had they seen them, they could not have interpreted them. For example, the committee wrote of Christian Kloc that "the field of measurements where data manipulation and misrepresentation took place is outside his specific expertise. It would be unreasonable to expect him to notice data misrepresentations that were for so long missed by even experts in the field" (Investigation Report 2002, p. 17). Note already the mediation implicit to this account: the committee did not even question why Kloc never witnessed the data generation but cleared him because he was incapable of a critical stance with regard to another, different layer of mediation in which data were (mis)represented. Zhenan Bao was likewise judged to have "met reasonably her responsibilities as a coauthor, for reasons basically similar to those in the case of Christian Kloc" (Investigation Report 2002, p. 18) – despite the fact that she was the corresponding author on one of the retracted papers. Authorship of a paper thus apparently does not require that one understand or be able to understand its fundamental claims or – which amounts to the same thing – the evidence for those claims. Sufficient unto the day instead is a passing fluency (perhaps little more than a circumstantially useful pidgin) with particular modes of representation many steps removed from the supposed "confrontation with reality" extolled by Sokal and, not presented, but represented by the initial experiment. That modern science is comfortable with such a low standard of evidence is suggested by, e.g., the American Physical Society's explicit revision of its ethical guidelines, following the Schön affair, to state that coauthors are not equally responsible for a paper (Brumfiel 2002c). In a 2007 editorial, *Nature* followed suit: "It is too glib to state that every coauthor of a paper shares full responsibility for its content. A researcher who specializes in the radioactive dating of rock strata cannot necessarily be expected to vouch for a paleontologist's analysis of the fossils within them – especially if the work has been carried out in labs on different continents." It is hard to imagine Irving Langmuir or R. W. Wood quickly agreeing.

A final and revealing expression of this logic of authorial non-responsibility came in the 2003 retraction of Schön's seven *Nature* papers. Only five of the seven had been scrutinized and condemned by the Investigation Committee. The other two had neither been challenged nor "found… formally lacking," in the words of the corresponding *Nature* editorial. Other than Schön's suddenly unsavory reputation, no evidence called the papers into question. So why were these two papers retracted? According to some of the retraction notices, it was in part because the coauthors could not "vouch for the validity of the data" (e.g. Schön, Kloc and Batlogg 2003; Schön, Kloc, Bucher and Batlogg 2003). But the reason why they were "completely cleared" was precisely because *at no point* were they in a position to vouch for the validity of the data. How can this simultaneously be a legitimate reason for retraction and an acceptable characteristic of coauthors? Yet, far from raising such questions, the corresponding *Nature* editorial instead commended the coauthors for the extra retractions (which *Nature* insisted it did not demand, and which Schön did not join), noting that "[Schön's] former colleagues and collaborators have now assumed in full their scientific responsibilities as coauthors on the papers in question." How extraordinary! The coauthors once again signed their names to a series of papers (the retractions) without demanding basic evidence to support their position, other than the metonymic, highly mediated evidence represented by the renown or now-infamy of the lead author. If this is *Nature*'s view of full scientific responsibility, how is it different from the prior behavior of these coauthors that *Nature*'s commendation implicitly criticizes? And what does such a shallow if not empty notion of "full responsibility" reveal about the present scientific enterprise?

As the beginning of an answer to this question, we can consider the hypothesis that the modern predilection to "crusade against scientific fraud" reveals "the unadventurous nature of scientists today, and their failure to understand the nature of science – to appreciate that judgment is required in assessing the validity of scientific claims" (James 1995). James went on to argue that at issue ultimately is a desire to adjudicate proposed knowledge on the basis of trust, rather than judgment, even though "trust should not be sought in science." "It is a sign of the current intellectual malaise," he wrote, "that it should be so widely yearned for." Much about this diagnosis is correct. Certainly, it was trust if only in authority which Paul Grant (2002) invoked in the early stages of the Schön drama: "[T]his is Bell Labs, and although these guys were my scientific adversaries for many years, I have the highest respect for their competence, credibility and, indeed, collegiality, and will accept their claims until proven otherwise." Likewise Bertram Batlogg: "I have placed, in retrospect, too much trust in my highly talented collaborator" (as quoted in Brumfiel 2002b). Schön's other coauthors also evidently relied on trust, first in his renown, then in his infamy. The Investigation Committee accepted such trust as an obvious corollary of modern science: "Collaborative scientific research requires a high level of trust among the participants" (Investigation Report 2002, p. 17). Even the editors of *Nature* (2002) accepted the evidentiary weight of trust: while "unequivocally reject[ing]" the charges of several critics (including high profile figures like the Nobelists Robert Laughlin [Adam and Knight 2002] and Philip Anderson [Brumfiel 2002b]) that *Nature* (and *Science*) had rigged their peer-review systems to facilitate publishing sexy findings like Schön's, they went on in the same article nonetheless to equivocate that "there is always the possibility that exciting papers from groups with excellent track records will be viewed less critically than those from an unknown quantity, as referees and editors might take more on trust..." The emptiness of this exculpatory equivocation is revealed in a simple question: for the first years of the millennium, who had a better track record than Schön?[9]

[9] To this may be added the results of a survey of 1576 researchers conducted by *Nature* specifically concerning a possible 'crisis' in reproducibility (Baker and Penny 2016). While acknowledging that "[d]ata on how much of the scientific literature is reproducible are rare and generally bleak," the article went on to report that nonetheless "73% [of respondents] said that they think that at least half the papers in their field can be trusted..." But on what basis do they think this? And what exactly do they mean by "trust"? The

How far does trust extend? In 2002, *Nature* disowned a controversial paper that it had published the previous year when it emerged that the primary data on which the paper's claims were based were likely artifactual – and that neither reviewers nor editors had apparently bothered to look at them (Adam and Knight 2002). Other studies of peer review suggest that it is often nominal, occasionally arbitrary and sometimes fraudulent. Thus, in a hoax dwarfing Sokal's in scope and, unlike his, involving refereed journals not specializing in interventionism or opinion, Bohannon (2013) submitted a purposefully flawed paper with major, "high-school" level errors to 304 peer-reviewed, open-access scientific journals. 157 of them accepted it. Only 36 of the 304 generated reviews that explicitly noted any of the major scientific flaws. But the problem goes beyond open-access journals. Bohannon himself quoted University of Pennsylvania biologist David Roos as saying that he "strongly suspect[ed] you would get the same result" with traditional journals. Indeed, as part of a transparent peer-review test, Schroter et al. (2008) introduced 9 significant errors into papers examined by over 600 reviewers of a traditional journal; fewer than 3 were typically detected, even after special training. The already-mentioned fate of rejected manuscripts, the majority of which are simply published elsewhere, also supports, at best, highly inconsistent peer review. Such was the conclusion of Peters and Ceci (1982): when they resubmitted lightly altered versions of already published manuscripts to the same journals which had published them 18 – 32 months earlier, only 3 of 12 were

---

invocation here of trust is especially noteworthy since the question was not whether research was conducted or reported in good faith.

detected as resubmissions, while 8 of the remaining 9 were rejected, often for unspecified but serious "methodological flaws." Notably, among the light alterations introduced by Peters and Ceci was the replacement of prestigious institutional affiliations in the original papers with made-up ones, suggesting that reviewer decisions may have been affected by institutional authority. Outcomes such as these have led some to reject peer review altogether, as "collapsing under its own weight" and too easily subject to the whims and abuses of competing scientists (Vitek Tracz, as quoted by Rabesandratana 2013). Certainly, peer-review is open to other types of abuse as well, such as authors who propose close friends or even themselves as reviewers. Such peer-review rigging resulted in at least 110 retractions between 2012 and 2014 and underscores the dependence that editors often have on author-suggested reviewers (Ferguson, Marcus and Oransky 2014). In general, there is a frequently expressed sentiment that, despite peer-review, anything can get published; e.g. James (1995) expressed "little doubt" that he could publish a nonsense paper on the respective coital rates of right- and left-handed people.[10]

---

[10] To this litany of concerns about peer review may be added the suspicion that it tends to enforce a certain mechanization or formulaicism in thought, e.g. the often incoherent application, if not fetishization, of a statistical benchmark like $P < 0.05$, which may be demanded by reviewers or imposed by authors in anticipation of review. Such a suspicion, however, points to something deeper: for, on the one hand, we might wonder – and ought to wonder – whether thought can ever absolutely evade being mechanical or formulaic (discussed below); and, on the other, we might recognize as fundamental an inescapable rhetorical confusion. What, after all, is the rhetorical status of, e.g., the $P < 0.05$ benchmark, everywhere deployed but

In this context, we might contrast James' (1995) critique of unadventurous scientists who fail to realize that "judgment is required in assessing the validity of scientific claims" with *Science* editor-in-chief Marcia McNutt's (2013) explicit appreciation that "[t]he peer-review process provides some level of assurance about the accuracy of a study's conclusions, relieving each scientist from having to assess the veracity of each paper personally." This is the same thinking, or at least is continuous with the thinking, that underlies institutional reliance on publication metrics rather than article content to assess the merit of individual faculty. If a reader is relieved of having to assess the veracity of each paper, it's not more than a skip and a jump to relieving the reader of reading it altogether – as shall become clear below. Of course, McNutt doesn't understand that she is advocating the abrogation of scholarly responsibility and that she is replacing it ultimately with the rule of authority, based on, e.g., the prestige of the journal or

massively misunderstood? In effect, it is the same issue as that at stake when papers are cited without having been read or when metrics are relied upon to assess productivity, quality and content (discussed below). It is, in all of these cases, the issue of quotability, of the capacity to take something out of one context and place it in another, on the pretense that it functions the same in both. That is a *representational* pretense: the thing displaced, the quotation or the P-value, is taken to stand for the function it ostensibly fulfilled in the prior context. In thus being both de-contextualized and essentialized, it serves as a shorthand – or, if one prefers, a trope, a metonymy, synecdoche or metaphor. It also serves as a shortcut, a way of (as it were) getting to the future faster. While we might be tempted to think that the rhetoric is in the service of temporality, we could only think so were we convinced that temporality were not itself rhetorical. For more on this theme, see Chapter 1 as well as the Heideggerian critique at the end of this chapter.

of the author(s)[11] – even though among the roots of modern science is precisely a reaction against blind adherence to textual authority that supposedly an experimental confrontation with the reality of nature would displace. Yet if, in contrast to McNutt, James takes the high and more rigorous road of insisting on scientific judgment rather than trust, he nonetheless does so from a position of naïveté. For he neither asks whether trust is truly relinquishable nor why scientists find themselves in "the current intellectual malaise" where trust is "so widely yearned for." Like Bacon before him, he doesn't wonder why a restoration is (yet again) necessary.

After all, why is it that editors are so reliant on author-suggested reviewers? Why is it that reviewers are so prone to overlook mistakes? Why is it that coauthors can't be held responsible for the scientific papers to which they have signed their names? In fact, why is it that scientific papers now frequently have tens, hundreds or even thousands of authors (Hotz 2015; the current record, I believe, is 5154 authors – an alphabetized author list 24 pages long)? Most of the discussion of the rise of multi-authored papers has focused on how it impacts the distribution of credit; for example, Greene (2007) declared that the two "quantities" contained in a scientific paper were "the increment of new science and the credit for its discovery" and then spent the rest of his article considering the

[11] She also doesn't realize that that replacement of responsibility with authority will penetrate the very authority that it extols – such that prestigious journals with prestigious reviewers may well accept papers from prestigious authors, all based on an authority become disconnected from its source and ostensible justification. On more fundamental aspects of this representational problem, see Chapter 2's discussion of the imperfection of Darwin's abstract.

challenges to divvying out the latter: "Most [authors] will prefer full credit for a few papers to little or no credit for many, considering where it matters most: university committees in charge of tenure, promotion and salary increments based on scholarly production." But he, like most other commentators (for an exception, see Rennie and Flanagin 1994), takes for granted the legitimacy of the supposed "increment of new science" that has been produced, which is why he can so quickly focus on what matters most. This is itself symptomatic: why do he and his fellow commentators fail to ask, not only whether knowledge is quantifiable, much less whether it comes in increments, but how it is that we know that an "increment" of new science has been produced in the first place?[12] Among the most basic questions is this: if coauthors can't be expected to understand the papers that they ostensibly have helped to write, how can editors, reviewers or general readers be held to a higher standard? But if none of them is capable of critically assessing the claims of a paper, how do we know that it is in fact really knowledge? Who is it that knows the knowledge that a paper claims to report? And what counts as knowing?

Here is a place where the problems of reproducibility, (mis)representation and authorship converge. In an age of extraordinary scientific specialization and multidisciplinary

---

12 In their article describing the results of *Nature*'s survey on the reproducibility 'crisis,' Baker and Penny (2016) oddly argued that "[t]he challenge is not to eliminate problems with reproducibility in published work." Citing a biological psychologist, they clarified that "[b]eing at the cutting edge of science means that sometimes results will not be robust." But if the results are not robust or perhaps reproducible, how the devil does one know one is at the cutting edge of science?

collaboration, both of which typically aggrandize expertism and hence appeals to authority, knowledge would appear as the summation of sometimes quasi-infinitesimal contributions each of whose validities rests solely on the trust accorded their contributor. The synthesis itself is an interesting accomplishment, being the knowledge of a whole supposedly deduced from parts the manifold of which no single person knows. One might expect this nearly immaculately conceived knowledge to trouble those with high standards of evidence whose worldview demands that theories incessantly confront reality. Far from it; the epistemological status of the knowledge being produced is so little questioned that attempts to replicate it are unlikely while credit dispensation is prioritized. In such a world, the acceptance of the small contributions of knowledge that make possible a whole is based and can only be based on authority, trust or, dare one say it, faith; and where that is the criterion of knowing in one of the presumably most intimate endeavors of knowing, actual research, why would a higher standard be applied to the less intimate acceptance of someone else's published results? In short, the general lack of interest in reproducing findings is not simply the result of social headwinds that make funding and publication difficult. It is part and parcel of a passive and uncritical acceptance of results as true based on unavoidably metonymic criteria, notably the authority of an author or journal or the mere fact that it has been published. The same thing is at stake in the mindless application of statistics, "knowledge" of which is often confined to a software package, or analogously in the use of black-box instrumentation. It even pervades scientific pedagogy, which is routinely reduced to the transmission of facts accepted as such on the basis of a professor's or textbook's authority. Rather than decrying this situation as an unfortunate but no doubt accidental intellectual malaise, as James (1995) would have it, to be beaten back by redoubled vigilance and personal purity, we should instead ask what the essential necessity is that underlies these phenomena – phenomena which have as correlates ever-splintering specialization, a parallel rise in collaboration which is but the flip side of specialization, a fetishization but also impoverishment of authorship (including not only the proliferation of multiauthorship but also honorary, guest and ghost authorship; for dramatic examples, see Ross et al. 2008) and an abstention of readership, perhaps going even beyond what Marcia McNutt advocated (e.g., Simkin and Roychowdhury [2006] used modeling of misprinted citations to estimate that fewer than a quarter of scientists read the papers that they cite, while psychologist Dorothy Bishop [as quoted by Anonymous 2015b], based on her experiences as a reviewer, pointed out that "when a manuscript with 20+ authors has grammatical errors, typos and/or no page numbers, you wonder how many authors actually read it."). The logical outcome of this trajectory is clear – massive production (or even "mass production," as opposed to "discovery" or even "invention") of "knowledge" accepted on the basis of unquestioned authority and the truth of which no one is in a position to assess; a "knowledge," then, without a knower; and ultimately, like other forms of mass production, a complete mechanization of science so as to require no human judgment whatsoever.

And that is where we are explicitly headed. With "scientific discourse buckling at the seams" (Stone and Jasny 2013), increased attention is being focused on how to avoid the problem of judging and reading altogether. "Amid an ever-increasing mountain of research articles, data sets and other output," wrote the editors of *Nature* in 2013, "hard-pressed research funders and employers need shortcuts to identify and reward the work that really matters." In short, metonymies (metrics) are needed, in this case (as ultimately in all cases) metonymies of metonymies – metrics which somehow evaluate the metonymic re-presentations of some aspect of the supposed real world as claimed by a paper. Moreover, it is presumably not just the funders and employers who are hard-pressed and hence need these shortcuts; how is the individual scientist expected to perpetually crest this growing mountain while simultaneously maintaining his or her research productivity, i.e. commitment to progress? Acknowledging this problem, e.g. that "[s]earch engines and automated libraries will return more articles in response to a query than anyone has time to read," Waltz and Buchanan (2009) looked forward to the day when computer programs will have "the intelligence to read and interpret the online information for us." "These [robot] assistants are becoming not only faster but much broader in their capabilities – more knowledgeable, more creative and more self-reflective." In what sense the word "knowledgeable" (much less "self"!) is used here is unclear, unless knowledge is simply understood as a collection of information divorced from any criterion of belief or, for that matter, from a believer, the strong sense of which, Stephen Dedalus once pointed out, is the only sense. And why not?

Evans and Rzhetsky (2010), conceding the inability of human beings to keep up with the ongoing and ever-accelerating "explosion of knowledge,"[13] argued that in the future "[n]ew knowledge will simply emerge from mechanical application of algorithms that mine data for plausible patterns." Is this view of knowledge not utterly continuous with the dilution of authorship, responsibility and judgment – not to mention wonder – described above, as well as with the divinatory mining of data at the statistical altar?

Yet even the budding emergence of this brave new world is not enough. Whether new "knowledge" is generated by human being or computer algorithm (for example, by the robot scientist "Adam" that King et al. [2009] designed to perform, "capture and digitally curate all aspects of the scientific process," including hypothesis-generation, experimentation and experimental analysis), a method is required to assess its impact. But it is no longer sufficient to determine, albeit using questionable metrics (e.g. Abbott et al. 2010, van Noorden 2010), the impact that an article describing purported new knowledge has had; now it is necessary to predict that impact in advance. Wang, Song and Barabási (2013), in a paper considered of such general import that it was published in *Science*, developed precisely such a metric, one that "captures the total number of citations a paper will ever acquire," i.e. "the discovery's impact." That citations and even impact may have little to do with quality or truth – as suggested by the pessimist-universe interpretation of Rzhetsky et al. (2006) and the failures to replicate even highly-cited work

---

[13] Is this not well-put? Is not knowledge in an important sense exploding?

(e.g., Ioannidis et al. 2009, Prinz, Schlange and Asadullah 2011, Begley and Ellis 2012), as well as, in a different idiom, by Plato – matters as little as those premature scientific discoveries so far in advance of their times as to be wholly unappreciated by them (for example, Mendel's work; see also Barber 1961). Setting both those problems aside, as well as the problems of a self-fulfilling prophecy or a metric which people learn to game, Evans (2013) commented positively on Wang et al.'s result, noting that "[w]idespread, consistent prediction of an idea's future impact will necessarily speed the resolution of which ideas win and which lose in the competition for attention and resources." Unquestioned is whether winning ideas are necessarily true ideas or good ideas. Also unquestioned is the self-evident good of speeding things up. No argument is necessary to explain what the benefit of yet a further acceleration is. This despite the fact that motivating the perceived need for metrics, shortcuts and robot scientists is precisely every human being's inability to keep up in the first place. Yet paradoxically there is no need to keep up; the object more clearly in view after Wang et al. (2013) is, according to Evans (2013), "[t]he ability to automatically extract scientific claims from research articles and reason across them [which] should lead to the prediction or computational generation of promising new hypotheses… In this way, citation prediction represents one step on the path to creating algorithmic or robot 'scientists' that are more creative, risky, persistent, and wide-reading than ourselves."[14]

[14] Should we ask what kind of risks are acceptable?

In short, we should throw ourselves into the wave. If, per the example of Schön's coauthors, ours is not to wonder why, neither is it – or at least neither will it be – seemingly for us to know or to seek to know. While such abstention might appear a remarkable development in the progress of what is, after all, called science (from Latin *scientia*, "knowledge"), its actual continuity with the scientific tradition is perhaps the more shocking. In a mild form, Lambdin's (2012) critique of the use of statistics in science already points to a rather low-tech form of this mechanization: "[S]ignificance testing creates in the mind of researchers the impression of automating the difficult process of thinking through inferences, seemingly reducing the complex notion of scientific support to the mindless task of an assembly line inspector, stamping 'accept' or 'reject' on every good that is rolled along."[15] Even this is a rather more

---

[15] In saying this he was echoing Ronald Fisher's earlier (1956) critique: "The concept that the scientific worker can regard himself as an inert item in a vast co-operative concern working according to accepted rules, is encouraged by directing attention away from his duty to form correct scientific conclusions… and by stressing his supposed duty to mechanically make a succession of automatic 'decisions'… The idea that this responsibility can be delegated to a giant computer programmed with Decision Functions belongs to a phantasy…" (as quoted by Goodman 2016). He likewise railed against a mechanical acceptance procedure in his 1955 paper: "In an acceptance procedure… acceptance is irreversible, whether the evidence for it was strong or weak. It is the result of applying mechanically rules laid down in advance; no *thought* is given to the particular case, and the tester's state of mind, or his capacity for *learning*, is inoperative. By contrast, the conclusions drawn by a scientific worker from a test of significance are *provisional*, and involve an intelligent attempt to *understand* the experimental situation" (pp. 73 – 74, italics his). He concluded this paper with another appeal to duty: "We have the duty of formulating, of summarizing, and of communicating our conclusions, in intelligible form, in recognition of the right of *other* free

ennobled task for human beings than that envisioned by King et al. (2009): they touted that their robot-scientist Adam required only that a (human) technician provide consumables and remove waste. Voila, the god-like, Adam-creating human is at once also the lowliest galley slave.[16] But if this servile divinity or divine servility seems paradoxical, its roots remain traceably Baconian (as well as Christian): not only in the confusion of master and slave implicit to his project of mastering Nature by obeying her, thereby obediently becoming as god, but in precisely his method of so doing. In the epistle dedicatory to King James I preceding "The Great Instauration," Bacon promised to "have provided the machine" whereby "philosophy and the sciences may no longer float in air, but rest on the solid foundation of experience of every

---

minds to utilize them in making *their own* decisions." I sympathize with Fisher's concern, and yet the distinction he would make between mechanical application and thought seems far less sharp to me than he would maintain; and this is already indicated, first, by Fisher's apparent ignorance that this delegation of thought to mechanism corresponds to the earliest articulations of a Baconian project, as I shall discuss shortly; second, by the necessity of communicating conclusions in intelligible form – that is, in a language and according to a grammar whose rules have been laid down in advance; and third, by the invocation of duty. What is duty, what is its hold on us, and why is it necessary to invoke here? If we concede that a decision is not the result of a mechanical calculation – that instead a decision entails our own responsibility and is accordingly without ultimate alibi – it nonetheless does not follow that we can be certain a decision has been made, whether by ourselves or others, rather than being the secret result of, say, a mechanical process, grammatical necessity or inherited duty. If, anticipating the end of this essay, we were to frame this in Heideggerian terms, the question would – and will – always be, can I truly distinguish inauthentic from authentic Dasein?

[16] Cf. Parodos, pp. 14 - 15.

kind." In the preface to The New Organon he resumed the metaphor: "There remains but one course for the recovery of a sound and healthy condition – namely, that the entire work of the understanding be commenced afresh, and the mind itself be from the very outset not left to take its own course, but guided at every step; and the business be done as if by machinery." It is unclear whether Bacon realized that what began as a metaphor or a dream of our imagination[17] may very well also become a pattern of the world – that, whether God forbid it or not, giving the one out for the other is no less the achievement of his method than what his method would prevent. In short, what began as if a machine may end as a machine. Such would be one form of the crisis of modern science – but only if machines are capable of crisis.

Part Two: The metonymies of men

> πάντες ἄνθρωποι τοῦ εἰδέναι ὀρέγονται φύσει.
> All men by nature desire to know.
> Aristotle, translated by W. D. Ross

So famously begins Aristotle's Metaphysics. At first glance, the claim appears both dogmatic and, at least in the modern world, mistaken. After all, how can such a desire to know be squared with a scientific endeavor that culminates in an abstention from human judgment, a knowledge without a human knower, or a knowledge contained within, or

[17] I am alluding to the quotation of Bacon on pp. 276 - 277.

(somehow, mysteriously) accessible only through, automata? If all men desire to know, surely they would not be satisfied to be authors on a paper only a small part of which they deeply understood. Surely they would not look forward to a day when their computers or robots did their reading, hypothesis-generation, experiments and analyses for them. Surely they would not be tempted to massage, much less fabricate, data, or passively to submit their data to the opaque arbitration of a statistical software package.

But perhaps in writing all of this, perhaps in being so sure, I am assuming too much. What does it mean "to know"? Do I know something if I have it on the spoken or written authority of a respected scientist, or for that matter, of an ancient philosopher? Or if it is the output of a supercomputer, a calculator, a slide-rule, the rules of algebra or the rules of grammar? Is knowledge dependent ultimately on being brought, in practice as well as in principle, to some sort of original self-evidence, from which everything else follows – that foundational moment of apodictic certainty which commands assent from all human beings, whether Thessalian nobility or mere slave boys, whether Euclid, Lobachevsky or Riemann? Must we, like Descartes, "raze everything to the ground and begin again from the original foundations… [in order] to establish anything firm and lasting in the sciences" (First Meditation)? But on what basis do we know that there is an original foundation? Descartes never explicitly asked this question; he treated as self-evident that an original foundation existed, even if we are always thrown into a world far removed from it. Given that circumstance, however, it falls to us to establish that original foundation, in the many senses of that verb. Descartes went about such establishment, establishment which would also be an excavation, by asking what his ultimate criterion was by which to know something as true. His solution, hardly satisfying, is instructive: at the end of his Fourth Discourse, he argued that what we know as true is what we are able to think clearly and distinctly; this is guaranteed by a God whose goodness precludes deception, and whose existence and perfection are known precisely because thought clearly and distinctly. Is the foundation of knowledge to be found(ed) in a circular argument? If so, then this inquiry, far from being begun by a mind "quite free from prejudices," as he described the necessary condition of his investigation in the letter dedicatory of the Meditations, relied on a primordial prejudice, as it were an originary pre-judging, as knowledge's Atlas-like pedestal. We risk – perhaps unavoidably – having answered by fiat the question "How do we know that we know?" precisely because for us the question demands an origin, as much stopping as starting place, that can only justify itself by asserting itself.

In approaching this problem we cannot help but circle as well an ill-reputed paradox attributed to the Thessalian student of Gorgias, Meno, and found in Plato's dialogue of that name. How do you seek, and how will you know that you have found, what you do not know? Socrates famously treated the question as eristical, as a rhetorical, sophistical and all-too-familiar threat to the very possibility of inquiry, for which latter we would have either no need or no capacity. (The familiarity of this threat already suggests that our desire to know is not unadulterated, or at least not simply reducible to a parallel desire to inquire.) Relying on the authority of various priests and priestesses and the poet Pindar, Socrates parried Meno's challenge by claiming that all knowledge was recollection made possible by the deathlessness and prior experiences of the soul. The eureka moment of intellectual discovery is quite literally a recognition, a re-cognition. This, however, is to resolve a problem with an answer that is just the problem in disguise: after all, if I know something now because I recognize it, having seen it in a past life, how did I know it then or for the first time? The risk is clearly of infinite regression, recollection like turtles all the way down. As with Descartes, we again confront a problem of origin and of end: an inquiry into the possibility of knowledge can stop only if a beginning is found(ed). And, again as with Descartes, the argument resorts to circularity: at 81c-d, knowledge as recollection follows from the asserted deathlessness of the soul; but at 86b, the deathlessness of the soul is inferred from all knowledge being recollection. That Socrates was aware that the question had been inadequately answered is explicit in the *Meno* – not least in his rigorous dressing of the benefits of inquiry in an optative rather than indicative mood (86b) – and is suggested by the return of the question in other dialogues and especially in the *Theaetetus*. There an aged Socrates, on his way to the King's Porch to answer, unpersuasively, the charges against him, encountered a young mathematician and the ghost of Protagoras and could only reveal, time after time, how their answers to the question of what knowledge is were inadequate, mere wind-eggs. Not incidentally, our account of this later engagement is, like Protagoras, strangely spectral, third-hand, ostensibly the report by Plato of another written account read aloud by a slave to two men long after the events being recounted and while Theaetetus, the once-young mathematician, is at that very moment dying off-stage. The presumed original is thus obscured in a strictly unrecoverable, already distant and rapidly receding time. Like Descartes looking out his window at jostling hats and cloaks, we might worry that in our efforts to ground knowledge we have placed our trust only in feigned men, in specters, or, by another name, in automata.[18]

Thus Aristotle's claim that all men desire to know is mysterious if only because it is unclear whether we know what it means to know. In fact, however, the entire statement is problematic, becoming the more puzzling the more we reflect on it. That strangeness emerges in greater starkness if we forsake the familiar translation of Ross (and many others) for one more awkwardly literal: "All human beings, from a source of motion and of rest inherent to them, stretch themselves out to have seen." I would like to comment on three words in this statement, the last three of the quotation in Greek: εἰδέναι, ὀρέγονται and φύσει. The first is a perfect active infinitive usually translated as "to know" but more literally meaning "to have seen." Just as in English we may nod our heads and say "I see what you mean," so in Greek, to know something is to have seen it. Indeed, this meaning was very much on Aristotle's mind, as only a few lines after his opening sentence, he addressed this priority of sight, writing of it that "this, most

---

18 The 1647 French translation of The Meditations by the Duc de Luynes translates the Latin "automata" (in the quotation from the second Meditation with which I preface Part One of this chapter) as "des spectres ou des hommes feints qui ne se remuent que par ressorts" – "specters or feigned men who only move by springs."

of all the senses, makes us know" (γνωρίζειν). The second is a present middle-passive verb that certainly can connote desire, as Ross translated it, but has a more concrete sense of "stretching oneself out" or "being stretched out." It is a grasping or a reaching for, moreover, in this form, of ambiguous voice – either middle or passive. Finally, the third, translated usually as "nature," begs the question of what "nature" means to Aristotle. It is, according to his study of nature (The Physics), "a source or cause of being moved and of being at rest in that to which it belongs primarily, in virtue of itself and not in virtue of a concomitant attribute" (Book 2, ~192b20, translated by Hardie and Gaye).

Now we could certainly fall into a philological regress, exactly paralleling our pursuit of the foundation of knowledge, and further recapitulated in the mysterious source (archē, ἀρχη) at issue in "nature," were we to declare that these supposedly more literal renderings thereby gained a closer approach to the original meaning. In lieu of that dangerous claim, which can only still what instead must throb, I wish more modestly to propose my translation as a way of sensing what is latent but unfamiliar in a claim that might otherwise not seem radical. "All men by nature desire to know" – we might affirm or deny that proposition without suspecting its strangeness or its stakes. But: "All human beings, from a source of motion and of rest inherent to them, stretch themselves out to have seen" – is more puzzling. What is enigmatic in it unfolds itself along a tri-partite axis of interpretative and ultimately complementary questions. First, how does one stretch oneself out to have seen? Certainly I can imagine straining in order to see – but how do I extend myself, or how am I extended, by a seeing that is already in some way completed? The mystery then shifts from how an already completed action can motivate me nonetheless to stretch myself out towards its completion (the first question), instead to a second question: What can that stretching out possibly accomplish? How does it accomplish it? And finally, third, how does this stretching out to have seen cohere with my nature, with a source of motion and of rest that is within me? Indeed, is this stretching out to have seen a form of rest, like a taut bow, or a form of motion, like the expanding universe? Given the middle-passive ambiguity of the governing verb, we can relate this question as well to whether the stretching out is done by me, to me, or (somehow) both.

Let me propose that many of the same concerns that appeared piece-meal in the first part of this essay are at stake in this fraught or at least cumbersome translation of Aristotle's opening line in the Metaphysics. How does one stretch oneself out to have seen? One reports on the vision that would retrospectively justify the stretching, even if that vision was not, in any simple sense, actually seen. My contention here, however, is not one of dishonesty and certainly not one of historical revision; rather I am attempting something more abstract and subtle. The data, usually first mathematized, are selected, shaped and beautified, all in accordance with a "perfect story" which they tell and, simultaneously, by which they are determined in advance. It is whatever constitutes the perfection, the completion, of that story, as much as the story itself, that figures in the perfect tense "to have seen," and that entrains in its (not our) service the flexible mediations of re-presentation. In other words, those views that, in a necessarily partial and obscure way, we actually do see must become the (rhetorical) figure for that prior and already completed seeing. The latter mediates them further, just as this perfect story that both pre-exists and comes into existence through the data is, in this doubleness, also stuttered, also re-presented. The child is once again father to the man and, as with Plato and Descartes, what we know we know circularly, having known it before. Heidegger, in "The Age of World Picture," discusses this problem in a different idiom, the meaning to the ancient Greeks of τα μαθηματα (ta mathēmata) as a knowing-in-advance which he takes to be the essence of mathematics. The world and the data in some way de-scribing the world are mathematizable precisely and only because known in advance – because, as it were, the story is always already perfect and the seeing has always already been completed before we stretch ourselves out. In writing this, I am not pretending that all the miscellaneous details are known as such in advance; what is known in advance is rather the way in which the world is knowable, the form in which knowledge would be present, and the type of story that can assemble and, as it were, make present the miscellany of facts. In the words of Heidegger, again from "The Age of World Picture" (p. 119), a fixed ground plan – a foundation, a type of origin – is projected onto natural events and "[o]nly within the perspective of this ground plan does an event in nature become visible as such an event. This projected plan of nature finds its guarantee in the fact that physical research, in every one of its questioning steps, is bound in advance to adhere to it."

And yet we stretch ourselves out or are stretched out. What can that stretching out possibly accomplish, and how?

To see what has already been seen is strictly not possible; we cannot catch up with what is done. And yet that is precisely what we attempt, in an almost traumatic spirit: we endeavor to stretch, to catch up, by re-collecting, re-peating, re-plicating and re-producing. The perfect story must nonetheless be re-ported; a bygone age must be re-stored. In this observation emerges the commonality of two seemingly disparate phenomena: the determinative sway accorded to replicability, on the one hand, and on the other the frankly neurotic proliferative drive to reproduce through publication. In declaring that these two phenomena have a common basis I am trying to suggest that they are actually aspects of the same thing, and that the replication of experimental results has largely been displaced by a different mode of replication, publication, that is falsely taken as equivalent to it, a mistake that is also insightful for recognizing a secret kinship. That false equivalence is analogous as well to the confusion of idealized (or fabricated) representations of data for that of which they would be the representations, a substitution all the more inevitable since the completed seeing is always already inaccessible. Thus, just as Descartes found an origin that he might instead only have founded, so too do we ever confuse as present what is only re-presented. The completed seeing, as soon as it is repeated yet again, invites a reaching out insofar as that repetition only demands another: it displaces onto a further horizon what is still an already completed seeing in need of being reproduced – in need, in fact, of being seen for the first time again. The ideology or idée fixe of perpetual progress is a further expression of this reaching out that, far from being satiated by a completed seeing, is instead incessantly incited by it. Precisely because the already-completed-seeing can only be represented, rather than presented, it falls into an inexhaustible regress of re-presentation, where each representation, being incomplete, being faithful only partially to its hypothetical and in any case receding, ungraspable and already-gone referent, yields its place to a new representation no less, although differently, incomplete.[19] Progress, like the *perfectibilité* of Rousseau, is thereby a horizon constituted by the indefinite rhetorical potential of language. For the same reason we, who are only ever thrown into the world, can never catch up with an origin or find a *tabula rasa*.

These answers to this second question can be allegorized in the figure of the tourist – a figure already closely akin to the modern scientist, as the rise of scientific tourism might suggest. The tourist is neither content to have seen – he must

[19] Is this not reminiscent of the dynamic already discussed in Chapter 2, e.g., with regard to "place"? And yet, would that in any way be surprising? It is also reminiscent, one more time, of the description of the man of science, Aylmer, in Nathaniel Hawthorne's "The Birthmark," at least as he is perceived by his wife whose unsettling birthmark he is endeavoring to remove: "Her heart exulted, while it trembled, at his honorable love – so pure and lofty that it would accept nothing less than perfection nor miserably make itself contented with an earthlier nature than he had dreamed of. She felt how much more precious was such a sentiment than that meaner kind which would have borne with the imperfection [of her birthmark] for her sake, and have been guilty of treason to holy love by degrading its perfect idea to the level of the actual; and with her whole spirit she prayed that, for a single moment, she might satisfy his highest and deepest conception. Longer than one moment she well knew it could not be; for his spirit was ever on the march, ever ascending, and each instant required something that was beyond the scope of the instant before."

photograph in order to see again and in order for others, through him, to see and, as it were, reproduce his findings – nor content, having seen, to stay put, but immediately prepares his next trip, his next viewing. There is no time for dwelling in the completed seeing, a consequence not of some spiritual shortcoming, his failure to "be in the moment," but rather of its always already having been completed. Meanwhile, pioneering work beckons, even at the expense of revisiting or replicating results and even as a photograph or a publication is still necessary to reveal, or rather promise, that indeed a seeing was completed, was captured, was in principle repeatable. That a seeing, always already completed, was nonetheless, if only for an instant, present – since, after all, here it is, re-presented. Yet, far from being a picture of the world, this representation is rather the "world conceived and grasped as picture" (Heidegger, p. 129).

How then does this stretching out to have seen cohere with human nature, with a source of motion and of rest that is within us? Here two points must be made. First, this source of motion, nature, is in Aristotle's opening metaphysical salvo paired to a verb of middle-passive voice; we – all human beings – are both the subject and the object of the action. Yet, whether we do it to ourselves (middle voice) or have it done to us (passive voice), our human stretching is nonetheless the accomplishment of our very own nature, which is grammatically its instrument.[20] In this way, our nature accommodates both our activity and passivity, whether or not we can distinguish them, while it brokers what is necessarily a

[20] Cf. Darwin's qualification of method with the unconscious and the related importance of the middle voice discussed in Chapter 3.

relationship, hence a mediation, with the completed seeing. The terms of that relationship are to be found internally, not externally – in our nature, not the completed seeing; even while that internality is not unambiguously that of either subject or object. Our nature may yet be alien while being our own. In science, a manifestation of this problem comes into focus with authorship, the apparent claim of responsibility, cause or source so important to the dispensation of credit and yet, by scientific convention, almost always articulated in the passive rather than active voice, now explicitly tied to a qualified responsibility up to and including a total lack of responsibility, and also increasingly parceled out in an indefinite division that can only undercut its assertion of atomism. Authorship in science is thus effectively of indeterminate voice, asserting contradictory claims of activity and passivity, and is the accomplishment of an often-lauded subject not easily, and not ultimately, distinguishable from an object.

Nor is it clear that it is an accomplishment of motion or of rest. This is the second point: in the same way that we may fall short of an absolute distinction between subject and object, or between self and other, so too may we fail to resolve whether our nature has driven us to move or to stay still. The scientist ever-pursuing pioneering work on the horizon, but to which he always has the same, determined-in-advance relation – is that scientist in motion or at rest?[21] Likewise the tourist,

[21] For example, the scientist "at the cutting edge" despite results which are far from robust, or the Galápagos scientists mentioned in Chapter 3 who have no doubt about the devastating role of rodents despite the absence of systematic study.

who always encounters the same other and the same relationship to the other, no matter where she is – has she moved or is she still? What kind of motion is it in which every confrontation with the world resolves itself once again into a completed seeing, once again into the same perfect but also programmatic story that, in the case of the tourist, inspired the trip in the first place?[22] Productivity understood in this way has as its correlate the busy inactivity of the Heideggerian "They" [*Das Man*], an inauthentic mode of Being that in our average everydayness, for the most part, is ours – but that, for all its ongoing activity and sheer commotion, has, as its condition, stasis above all inert before the overwhelming potential for being and, importantly, for not being that also is ours.

In short, the equally necessary aestheticization of data and perfection of stories (for which Schön's example might serve as aptly-named fable), the unfulfilled need for replication paired to exponentially increasing publication, and the emptying-out of authorship joined to a proliferation of authors, may all be coherent expressions of man's natural desire to know. Aristotle's claim faithfully describes the modern scientific project even in its pathological mode, although the aptness of that description is recognizable only on the condition of its strangeness. The desire to know, the stretching out to have seen, is strange. So long as that strangeness is obscured by the commonplace, for example by not inquiring into what it means to know, we risk, regarding the applicability of Aristotle's claim to modern science, an equally empty

---

[22] In Chapter 3, this was discussed in terms of an encounter with nature fitted to an aesthetic preconception of it

affirmation or denial: i.e., an affirmation which, without pausing, accepts that of course science is continually accreting knowledge, taken as self-evidently equivalent to information, and which is in no way troubled by the litany of concerns raised in the first part of this essay, each one of which is a mere accident as well as the price of progress; or a denial which moralistically denounces the present-day abrogation of scientific judgment but always against a backdrop of a golden era, the assumed existence of which renders an inquiry into the necessity of that abrogation pointless, since that abrogation is but the unfortunate, contingent and above all accidental outcome of history expressed in its, e.g., psychological or sociological dimensions. My claim is altogether different: what I have (somewhat dangerously) called the "symptoms" of crisis in modern science collectively point to an intelligible pathology that is itself an expression of the desire to know, of the stretching out to have seen, in all of its strangeness.

In being thus, "pathological" science is no different from any other mode of scientific inquiry. The Aristotelian, metaphysical chord vibrates through modern science in all of its states, whether "diseased" or "healthy." The diagnostic metaphor is accordingly dangerous if it obscures this more fundamental commonality. Yet I have resorted to it because no opportunity for crisis or criticism will be perceived in a science understood as, for the most part, working properly. Certainly this is especially true in a modern ideological atmosphere in which science is accorded immense authority and a magical, indeed panaceal or anti-panaceal, status, both as remedy and as cause of problem after problem. (The pro-science ideologues emphasize the remedies, the anti-science ones the causes – but it is only on this relatively minor conclusion that they fundamentally disagree.) My attempt instead is to sidestep such ideology without merely opposing (and hence becoming) it, by showing science playing out, in deeds and in words to both of which it is overdeterminedly inattentive, an almost parody-like autocriticism. In this there is occasion for crisis and for more serious criticism. It is, after all, already there, if anyone would notice. But that opportunity would be entirely wasted if reduced to something curative, whether a faith in science's ability to self-correct or a more directed (and incidentally impossible) excision from science of its diseased tissue – in fact, the stances of the affirming and denying camps above. "Disease" is thus to be understood not as an anomalous but as an exaggerated state, one which provides the opportunity to understand science anew and differently veiled.

The science revealed in this re-veiling, however, is not particularly new. Heidegger predicted many of its aspects in the first half of the last century. He foresaw in 1938 the disappearance of the scholar and his replacement by a now familiar figure, "the research man who is engaged in research projects. These, rather than the cultivating of erudition, lend to his work its atmosphere of incisiveness. The research man no longer needs a library at home. Moreover he is constantly on the move. He negotiates at meetings and collects information at congresses. He contracts for commissions with publishers. The latter now determine along with him which books must be written" ("The Age of World Picture," p. 125). Compare that prediction to Peter Lawrence's (2002) description of the modern scientist in *Nature*: he is "swanning around the globe or running companies," attending a "conference circuit...

designed to build up a few stars," harnessing to his advantage "the exponential rise in the secondary literature" and the concomitant limitations journals impose on form and content (including their preference for positive results), and treating his students like technicians, "like boosters on space rockets... [which] when their fuel is spent, fall to the ground as burnt-out shells." For Lawrence, these are all examples of rank injustice and are thus in principle correctable. For Heidegger, they reflect a very specific comportment toward Being; to treat them simply as undesirable deviations is both to miss and to exemplify the point.

In "The Question Concerning Technology," Heidegger discussed that comportment in terms of the essence of modern technology. According to him, that essence "starts man upon the way of that revealing through which the real everywhere, more or less distinctly, becomes standing reserve" (p. 24). By "standing reserve" he was referring to how "[e]verywhere everything is ordered to stand by, to be immediately at hand, indeed to stand there just so that it may be on call for a further ordering" (p. 17). The Rhine, for example, becomes a form of energy storage or a stop on an organized tourist trip. Standing reserve also characterizes the prevailing view of knowledge as equivalent simply to information; knowledge as standing reserve is ordered as information, stock-piled, for example, in an online database and available as needed. We do not inhabit knowledge; we access it, withdraw it like currency from a bank, and it is accordingly hard, and perhaps not possible, to distinguish a scholar from a person with a smartphone. Common to these examples, whether of the Rhine as water power supplier and photographic backdrop or of knowledge as an information bank, is their appearance as at our command (p. 16). To be clear, this is not a transformation of the world into something we command. Standing reserve does not come about in answer to a considered preference that we may have for a particular way that the world appears. Rather it expresses a comportment toward Being that we find ourselves in and according to which the world appears to us. Other comportments may be possible, but to inventory them and await a selection is to understand them through the comportment of which standing reserve is an expression.

Modern science is in no way understandable apart from technology, from a way of revealing in which everything appears as standing reserve. Indeed, all of the problems in modern science identified in the first part of this essay follow from the fact that science can only encounter itself as it encounters anything real: as standing reserve. Think about it: in science, the physical world comes to appear to us as usually mathematized data, stored, for example, in Microsoft Excel spreadsheets on one's laptop or in the enormous, public data repositories proposed as the solution to data misrepresentation. There they stand by. Traditional and open-access journals, of which there are ever more, similarly appear as archives, whether or not they have readers, and stand by. Metrics are created to further order the order in which databases and archives store, and that new order stands by. Robots are built to comb those repositories and archives, to generate thereby hypotheses, and to test those hypotheses. Why? In order to contribute to the very same repositories and archives[23] and

[23] Indeed, catacombs.

there stand by. The robots require what information science calls an ontology, i.e. "a formal naming and definition of the types, properties and interrelationships of the entities that really or fundamentally exist for a particular domain of discourse" (Wikipedia). But what is this so-called practical application of philosophical ontology, if not an ordering for the purpose of standing by? Is the comportment implicit to standing reserve not already suggested when "ontology" is modified by an indefinite article? Not surprisingly, such an ontology can readily be deployed and is conveniently citable, as we might expect of standing reserve: "[W]e used the ontology of scientific experiments EXPO. This ontology formalizes generic knowledge about experiments" (King et al. 2009, describing the programming of robot scientist Adam). Even the future would be ordered to stand by, whether in the metric described earlier that promises to predict all the citations a paper will ever acquire, or in the massive, supercomputer-run simulations of future climate, or in the genetic prophecy of future illness or simply in science's veneration of prediction per se. Finally, the enormous interpenetration of modern science and modern capitalism – often decried as a conflict of interest and cited as a sociological factor contributing to questionable scientific practices or outright fraud – is itself but an expression of this comportment toward Being: for, like science, capitalism would order the world, including the future, to stand by. Where the one mathematizes, the other monetizes; the gestures are virtually identical. As expressions of the same comportment toward Being, their entanglement is thus no accident. Indeed, were we to cast our glance backward, we would find evidence of that entanglement as early as

Bacon's articulation of his method, whose benefits he linked to both commerce and the military, and whose procedure began with a laying out in tables of all the instances of a particular phenomenon, a cataloguing made possible only when the real can appear as standing reserve.

In all of these instances, from Paradise-restoring Bacon to the new robotic Adam, what appears appears as at our command. It is through the ordering of what's real to stand by, whether in tables or electronic databases, that human beings would achieve their dominion ultimately over the very universe, as Bacon ambitiously ordained in the 129th aphorism of Book 1 of The New Organon.[24] The aspirations of modern science are continuous with that ambition, as testified by projects, always both scientific and technological, which range from synthetic life, barcoding all living things and directed gene editing, on the one hand, to future Mars colonization, the deliberate geoengineering of Earth and eventual Dyson spheres, on the other. Even when the role of random effects and the sensitivity to (always mythical) initial conditions are recognized – as in the famous example of the butterfly effect –

[24] Moreover, this Baconian project (discussed also in Chapter 3) is often encountered in the remarkable mode of its somehow already having been accomplished: e.g., in a panel discussion on climate change and geoengineering, Harvard atmospheric chemist Jim Anderson identified our challenge as "regaining control of the [global climate] system" – a stunning statement if only for its use of the word "regaining." Yet in the context of my analysis of Aristotle's claim about the human relationship to knowledge, we may see a metaphysical necessity to the assertion of both completeness and necessary repetition (a re-gaining). (The online panel discussion was associated with the November 16th 2007 issue of *Science* magazine: http://www.sciencemag.org/cgi/content/full/318/5853/1054/DC1.)

the solution, for modern science, is simply to, in the words of MIT climatologist Kerry Emanuel, "tame this butterfly."[25] Butterflies, no more than chance events, cannot fail to be barcoded and ordered into standing reserve. In fact, science is incapable of encountering anything that resists that ordering. The resultant dominion is therefore a strange one. The inevitable accident is always understood retrospectively in terms of incomplete knowledge and hence, theoretically, was avoidable. The necessity, even essentiality, of accidents is precluded; no event, including those of catastrophic scope, fails to be ordered to stand by, ostensibly towards the completion of knowledge which would serve as a bulwark against the event happening again or, in a certain way, having happened at all. In this mode, man's dominion is over that with which man is ever trying to catch up – something which has always already been completed, something which we do not see and are not seeing but only have seen. It is a dominion which presides over a universe strangely inert, ordered, but ordered to stand by, warehoused or deposited, as it were, in a bank account, and accessible to us only through further orderings, orderings of orderings. Like the use of scientific metrics – whether in measurement or as proxies for reading articles – the structure thus described is metonymic. It is an illusion to think that we can get to the bottom of that ordering, its end or its beginning, because that bottom would not be an order but what is ordered and precedes order; but what appears to science appears as standing reserve, as already ordered. The god-like role in which modern science casts the

---

[25] Kerry Emanuel made this comment on August 1, 2007 on NPR's *Fresh Air*: http://www.npr.org/templates/story/story.php?storyId=12421331

human being is accordingly qualified: our dominion is over the inert, a completed seeing with which we cannot catch up, and further, which is accessible to us only through an indefinitely extending progression of metonymies which simultaneously infinitizes us – by rendering, seemingly without limits, more within our grasp – and isolates us – because what is delivered to our grasp is graspable only as mediated.[26]

Put otherwise, the condition of human dominion, of the appearance as at our command, and accordingly of the scientific project, is what Heidegger (in Being and Time) would call an inauthentic mode of Being, the They-self [*Das Man*] that in our average everydayness we are. The They-self is not truly individuated – through it "everyone is the other and no one is himself" (p. 165) – and can confront no unanswerable problem, not even death. For "'Death' is encountered as a well-known event occurring within-the-world" (p. 297). It too can be ordered as standing reserve, like any other event pretending toward – but for the They-self incapable of truly achieving – catastrophe. Death is a problem like any other for the They-self. Far from representing the possibility of the impossibility of existence, indeed, far from representing the absolute impossibility of existence, death, like any other problem, fails to be vanquished (which is to say, ordered) only to the extent that accidents keep happening and knowledge, though organized as a perfect story, is as yet, merely circumstantially, incomplete. This is no less the scientific account; indeed, modern science would be flummoxed by a claimed possibility of the impossibility of existence, since the

26 Thus, like Zeus, do we become all-wise (μητίετα) – but on the condition that wisdom (Μῆτις) is a fly

impossibility of existence is precluded by the fact of existence, nicely ordered and standing by. Neither for the They-self nor for science can death truly appear as an existential problem, one in which we confront our ownmost, non-relational, not-to-be-outstripped potential for being as necessarily including the omnipresent possibility of not being at all. Rather, science and the They-self take a 'practical' approach; death remains an eventual concern, something which can happen under abnormal, pathological or, in any event, exceptional circumstances, and to which efforts even now are being directed to establish appropriate treatments and cures, up to and necessarily including the project of dominion over the entire universe.

If modern science is capable of crisis, that crisis would be readable in the figure of the modern scientist – caricatured in *Nature* (as noted in the Parodos) as a superhero and in every respect, as such a cartoon might imply, simultaneously aggrandized and belittled, not unlike automata dressed as men or, indeed, ironmen. The modern scientist would be infinite, capable of both mastering and contributing to an exponentially increasing font of knowledge; but she is also so terribly finite, indeed so infinitesimally small, that she may not read the papers she cites, author the articles on which she is named or be held responsible for those articles' content. The modern scientist would swan around the globe, manage start-ups and lead a lab while publishing a paper per week – for him time is no obstacle; but he also hasn't the time to do the experiment himself or to witness its performance, much less to read critically, carefully referee a paper or attempt to replicate someone else's work. The modern scientist rides progress bareback like a bull at a rodeo but needs the future to be parameterized in metrics so as to know in advance what will have mattered – while mysteriously never wondering whether it is the rider or the bull who is in charge. In short, that towards which the modern scientist pretends is a human being who has surpassed all limits, a god-like human, a deathless human, a superhero; but this very pretense reveals a human being who is all-too-finite, all-too-dependent on trust, authority, software packages, dimly understood technologies and grad student booster rockets. Indeed, the human being finally infinitized is also a nullity. Yet this project of human infinitization and nullification can only reveal a being whose finitude is a concern to her and underlies everything that she does, even as it is always thematized as something else. It isn't just that modern science, following the lead of Bacon and Descartes, has established as its explicit goal the indefinite extension of human life and human power; the conduct of modern science is itself already an expression of that goal. To the degree that this goal is achieved, it will be as a standing reserve in which are ordered all real things, including not only science but the scientists themselves. Looking out our window and one last time taking a metonymy for the thing itself, we will properly wonder whether automata, specters and feigned men aren't simply what we fundamentally are and have always been.

But there will be also that which does not appear. What cannot appear as standing reserve is the possibility of the absolute impossibility of existence. This possibility cannot merely take its assigned place and wait. It is non-localizable. There is no principle by which it is ordered. It hovers about us in every direction, it will not be still and it cannot be tamed. No figure is adequate to it, and it can be approached only through figures.[27] I do not think modern science can face this possibility – a facing which the Heidegger of Being and Time would consider a more authentic mode of being. I think modern science is so constituted as to preclude even the suspicion of this possibility. Quite simply, despite its ravenous, ever-growing scope, modern science cannot encompass that possibility which orients our being, whether in authentic or inauthentic modes. In this sense, the crisis of modern science, the criticism pointing to its unsurpassable limitation, is nothing other than its incapacity for crisis. What would come were that to change?

What would come indeed. The simple answer is to imagine a science capable of comporting itself authentically with respect to our ownmost potentiality for being. But this simple answer is also simple-minded. Here the language of authenticity and inauthenticity may underwrite only a comfortable return to the philosophical presuppositions from which we began. Can science – in fact, can anything – face the possibility of the impossibility of existence? Or, more starkly, the impossibility of existence? That "face" is the problem, no less for the authentic than the inauthentic modes of Being; how are we to figure, and how are we to avoid figuring, what is not there? Will not the metonymies of men obscure that of

[27] Whether the figure of extinction which "almost inevitably" accompanies Darwin's theory of evolution by Natural Selection, or the (related) figure of gaps on which the geological record's coherence oddly relies, or the (related) figure of blankness against which are articulated letters, words and the semblances of meanings.

which there is no metonymy, no metaphor, no trope – though it is 'known' to us only as these? There will be no science of what is not; no other science will ever have been; no science will ever have been. Can we nonetheless imagine a science to come in which these last three clauses will merit being read and endured?

Imagining such a science will require us both to recuperate and to put into question the more general sense of science, of *scientia*, as knowledge. For, on the one hand, is there anything to suggest that the critique that I have proposed of modern science, in the narrow understanding of that word, does not also apply to what are traditionally and problematically termed the humanities, if perhaps in a less exaggerated form? Are their journals, open-access or traditional, not proliferating? Are their scholars not being scrutinized at the unexamined altars of productivity and progress? Is the fragmentation of supposed knowledge not driving them as well toward authoritarian and collaborative approaches to both research and pedagogy, approaches which closely parallel those of the sciences and which even fuel a, for the most part, sterile rapprochement of the 'two cultures'? In short, did Sokal's attack, however sloppy, unserious and little thought, truly fail to hit even a single nerve? Yet, on the other hand, do not the sciences (again, in the narrow understanding of that word) need the questions, the problems and the critical methods currently asylumed almost exclusively in the so-called humanities? A science to come cannot persist within a naïve theory of reading and a dream of immediacy, any more than it can with an instrumentalized and unexamined philosophy or a principle of responsibility which accepts willful blindness as exonerative.

Above all, neither the so-called humanities nor the so-called sciences can continue within an understanding of knowledge which ultimately assumes its foundation (origin, archē, ἀρχη) and its end (telos, τελος). Which is at last to confess that we have always thought knowledge within a metaphysics which privileges presence – whether a past presence from which we have accidentally and incidentally become alienated, a future presence towards which we strive and imagine as eventually our own, or a present presence which declares what it cannot describe, the immediacy of self-evidence. Whatever knowledge is, it has always been understood as that which escapes the indefinite relay of reference, whether it does so as ultimate origin, ultimate end or immutable truth.[28] But everything I have described suggests that knowledge is also precisely not that: it is, for us, unthinkable – despite our illusions to the contrary – outside of a future horizon which always recedes, whether in its demand for further replication, further prediction or further progress (this may also be what is at issue in the mostly unexamined glorification of knowledge "for its own sake"); it is, for us, unthinkable beyond an already completed seeing, a having-seen, that has yet to come to grips with the absence of an originary seeing; and it is, for us, unthinkable as an immediacy which a moment's reflection won't reveal as already mediated. The disinterest of Schön's colleagues in witnessing, in being present for, any of his measurements, represents, then, not

---

[28] It would be the corollary of the being I would be.

simply the shortcomings of fallible and ambitious human beings, but their remarkable if implicit acknowledgement that they could never have been present enough.[29] Their willingness nonetheless to share authorship with him is no less the same acknowledgement. And certainly, more broadly, it is far from a coincidence that the symptoms of crisis in science – now understood in the broadest possible sense – manifest themselves, over and over again, in the guise of writing, that dangerous supplement, in which, as Derrida has shown convincingly in Of Grammatology, a metaphysics of presence is deconstructed by a signified continually revealed as a signifier.

If the structure of knowledge which we are forced to confront cannot abide our theory of knowledge, we must displace the latter. Knowledge must be re-thought, if it is possible, and cut loose from its imagined anchor to what will always have been a transcendental signified, the one chord in the entire universe which resonates only in the enduring present, when signification would both begin and end. But this is not to announce tabula rasa or carte blanche, or to fiddle for the dance of an untethered relativism. These can only end by affirming the metaphysics of presence, albeit in a negative mode. Rather, it must be from within a re-thinking of knowledge that the responsibility signed for by a signature on an author page is considered – a responsibility that, far from being circumscribed by the limits of a claimed knowledge, begins and must begin not only at those limits, but beyond them.

---

[29] In the idiom of Chapter 1, at issue is a being which is not – quite.

Exodus (ἔξοδος):
The Swan of Tuonela and Mourning's Thinking[1]

Modern science is conducted from within a metaphysics of presence which privileges the subject as agent. The inevitable mediations which nonetheless intervene – the human reliance on inheritances like language and grammar, an interrupted consciousness fully transparent to no one, our finitude and embodiment – pass unnoticed or are treated as mere accidents, ultimately to be supplemented, then vanquished, by knowledge. This role accorded knowledge for the liberation of human beings into the full presence awaiting them continues a millennia-old philosophical tradition but, read against the grain, it also reveals the desire which has resonated at the core of that tradition and which has always preceded and framed knowledge: for knowledge as ultimate deliverance, assumed to be both eternal and unchanging, cannot help but postulate a knower as ultimate deliverer, a subject whose agency is at last unadulterated by ignorance either of the world or of himself. The metaphysical economy in which knowledge operates demands that eventual, unmediated knower: god in the beginning, man in the end.

This aspiration, this dream, would require the end of signification.[2] No more will it be in play; no more will one thing take the place of or stand for another. Otherwise such substitutions introduce an ineradicable equivocation: does this

---

[1] The Swan of Tuonela is a tone poem that is part of the Lemminkäinen Suite (Four Legends from the Kalevala) of Sibelius.

[2] This end, furthermore, will be imposed as either teleology or eschatology – nor will these two possibilities fail eventually to converge.

this stand for itself or for something else? The immediacy of knowledge would thereby be threatened, and with it the immediacy of the knower.[3] Instead, the dream is that our language, as Darwin promised, "will cease to be metaphorical, and will have a plain signification" (Chapter XIV, p. 485). But a language which ceases to be metaphorical will not signify; a language which relinquishes the possibility of metaphor is unthinkable. Such a language will have no common nouns, no species, no groupings of things more or less alike; it will devolve into absolute particularity, into a stuttering Tower-of-Babel babble. We must recognize that no category exists without mediation.[4] And we must recognize the corollary which results, that there is no immediate thought and that there can be no immediate thinker.

Yet we persist in this dream, nowhere more emphatically than in the modern scientific enterprise, which has borrowed against the dream's promised bounty. Our disturbing oneirism is hardly subtle when scientific superheroes strut and fret their

---

[3] This remains true even if we consider the role of knower relegated to a computer. Such a computer-knower will nonetheless rely on a so-called ontology (of the "applied" variety) which attempts to describe the world in absolutely unequivocal terms. But such terms are not available. The pretense to them is a manifestation of the metaphysics of presence insofar as it inscribes the articulator of the ontology in the role of absolute knower. That person must wield language with unadulterated mastery; language must be completely transparent to her in order to define unequivocally an equally transparent ontology according to which the computer would, again unequivocally, operate.

[4] It is important to bear in mind that these distinctions applied to the presumed but impossible language are distinctions imposed already by the language that I am in; from within that language the language I would describe can only be approached, as it were, catachrestically.

hours upon the stage. Yet the more probing critique may begin not with caricatures and costumes but rather with the characteristics attributed to scientific writing. This writing would pretend towards a language which is at once an analytical method ("The art of reasoning is nothing more than a language well-arranged," claimed the Abbé de Condillac as quoted by Lavoisier in the preface of The Elements of Chemistry) and the expression of an objectivity which it uncritically postulates and necessarily lacks.[5] On this

[5] This pursuit and its ready achievement for the most part are taken for granted, though occasionally scientists have foregrounded them as an integral part of their project – for example, in Lavoisier's appropriation of Condillac's critique of language in order to ground chemistry's foundations (in the preface of The Elements of Chemistry), first by a purgation of linguistic abuses always understood as unfortunate in their consequences and accidental in their cause, and second by a thereafter carefully controlled deployment of language whose inevitable 'degeneration' into what it corrects fails to be foreseen. In Lavoisier's case, as much as he tried to govern his language – his project began "from the necessity of reforming and completing the Nomenclature of Chemistry," as he stated in his preface, and took for granted, "as three impressions of the same seal," the separability of facts from their representation as ideas and of ideas from their expression in words – it was often his language which nonetheless governed him. For example, the composition of muriatic acid (now known to be composed of hydrogen and chlorine) remained for Lavoisier an unsolved mystery because of his linguistic association of acids with oxygen, the latter being a term he coined for an element which, he thought, generated acids. (Cf. p. 51 of The Elements: "We have given to the base of the former, or respirable portion of the air, the name of *oxygen*, from οξυς *acidum*, and γεινομαι, *gignor*; because, in reality, one of the most general properties of this base is to form acids, by combining with many different substances." Then, pages 71 – 2: "Although we have not yet been able, either to compose or to decompound this acid of sea-salt, we cannot have the smallest doubt that it, like all other acids, is composed by the union of oxygen with an acidifiable base. We have

understanding, language is simply deployed as a tool for the mere, unembroidered statement of fact. Moreover, the proper scientist employs this tool with complete mastery. Accordingly, metaphor and other rhetorical figures are strictly regulated if not forbidden outright, as if regulation or prohibition can by itself realize Darwin's prophecy. These measures, however, are not only impotent but always too late; their result, ultimately, is only that these tropes pass rigorously unnoticed and rigorously unread, often in ways that approach the Kafkaesque. Measurement and citation are thus exempted from their synecdochal affiliations, and to call DNA "software" is simply to acknowledge, syllogistically rather than figuratively, life's identity with computing machines. As for all those scientific authors – people who may not have written, read, understood or been able to understand the very words attributed to them and whose attribution they accepted –

---

therefore called this unknown substance the *muriatic base*, or *muriatic radical*, deriving this name, after the example of Mr Bergman and Mr de Morveau, from the Latin word *muria*, which was anciently used to signify sea-salt. Thus, without being able exactly to determine the component parts of *muriatic acid*, we design, by that term, a volatile acid, which retains the form of gas in the common temperature and pressure of our atmosphere, which combines with great facility, and in great quantity, with water, and whose acidifiable base adheres so very intimately with oxygen, that no method has hitherto been devised for separating them.") An analogous though more subtle example would be Lavoisier's deployment of the term "caloric." My point here is not to correct a past scientist from the light of present knowledge, but rather to observe that the sequential seal impression that Lavoisier postulated, grounded in fact, represented in ideas and properly expressed, if only care is taken, in words, assumes a linguistic propriety which has never existed, not least because words always already accompany every supposed fact – whether or not the fact is really there.

however much this approaches the rhetorical figure of apostrophe, however much this reveals how authors figure as figures, still the ultimate issue is simply the proper parceling out of a credit – as if this notion, credit, were elemental and foundational, atomizable in one-to-one correspondence with the supposed increments of knowledge, and not itself an implicit acknowledgement of presence deferred, presence as it were *credited* not despite but because of its absence.

In borrowing against a dream's promised bounty, however, science cannot help but critique its own metaphysical project. It leaves open that reading against the grain which I have already mentioned. What better parody is there of the unmediated knower come at last to the full possession of her plenitude than authorship lists twenty pages long and the ever-more-winnowed circumscription of authorial responsibility? Likewise, the scientist in control of evolution but unaware that his own project is driven by theological and mythical (not to mention metaphysical) discourses, or the self-sufficient cell, projection of our own desire, which nonetheless grades insensibly into the parasite, the organelle and the virus. What are these but unwitting satires and involuntary burlesques? The autocriticisms, in short, are available to be read. Precisely insofar as signification remains in play and its equivocations remain untamed, these autocriticisms can only tag alone like shadows or irritating mimes – though these latter are perhaps more substantive than that of which they are images. The problem will never be the absence of autocriticisms; the problem is instead always whether anyone will bother to read them. The scientific prohibition of rhetoric, whose roots extend at least to Bacon's promised exorcism of the four idols, has always been less a stricture on writing than on reading, beginning with the excessive simple-mindedness by which it recognizes, or fails to recognize, rhetoric in the first place. This is the very simple-mindedness which would also protect the city by expelling its poets, as if the city, the state, the leviathan, weren't already a poetic invention.[6]

Whether in a philosophical tradition stretching to the ancient Greeks, the Judeo-Christian tradition rooted in a garden, or in the foreshortened tradition of early modern science whose proximate foundation lies with Bacon and Descartes, knowledge has, again and again, been explicitly allied with the project of making man god-like. What, however, is this god whose likeness as well as image man would be? Would not this god be the author of a text exactly commensurate with his intention? Would not this god be the promise of meaning from within absolute plenitude, from within the purity of presence? Yet that commensurability and that promise betray a misunderstanding of meaning. For the possibility – not achievement – of meaning is conditioned by finitude, by imperfection, by knowledge, ability and presence everywhere limited. A philosophical tradition endeavoring to anchor meaning to an eternal, unchanging and singular Form has forever been mistaken. The condition of meaning's possibility – and equally of its impossibility – has instead always been what is transitory and changing, what is incapable

[6] It also parallels that simplemindedness which would return islands or even continents to an original, unsullied state, never suspecting that there is nothing which such a project will not ultimately have to recognize and treat as an impurity.

of quite being one with itself. If a sign means, it would be because it stands for what it is not; and this is true even when it stands for itself.[7] Were it otherwise we would not, for example, need scare quotes to announce when we were writing in such a way as to refer to a word in its wordness.[8] In all that is, there is always already an equivocation. Darwin's theory of evolution can be read as simply the elaboration of this single insight.[9]

To attribute plenitude to a god and then to pursue it for ourselves is to ignore as inessential the mediations already implicit in being the image or likeness of god, of being god-*like*;[10] it is, oddly, to forsake both the possibility *and* the impossibility of meaning. In a certain way this is a point made long ago in that scientific ode to atoms that Lucretius wisely

---

[7] As Shakespeare wrote in *MacBeth*, "Nothing is but what is not."

[8] Here one recalls Rousseau's brief account (in his "Essay on the origin of language") of the desirability of a diacritical sign marking irony. Far from resolving the ambiguities that always attend irony and the suspicion of irony, this mark would only compound them: for what precludes its possibly ironic use? More generally, one might observe that irony is the condition of possibility of meaning – even if it is also what always interrupts meaning and announces, as it were, meaning's simultaneous impossibility.

[9] Shall we say his is an insight into being, or into language? Perhaps into irony? Or rather is it an insight into writing?

[10] One could go further and observe that this gesture is to ignore the inessential *as inessential*; that is, the very distinction between essence and accident presumes a metaphysics of presence which has always been in a terrible hurry to rid itself of what is finite, limited, corporeal and unenduring – all those *mediations* which impress upon us our incapacity simply *to be* even as, without them, we are *not*. When Descartes stripped wax of its accidents, he was left with nothing recognizably waxen, just as cells stripped of their inessential genes end up gene- and genome-less – and eventually fail to be recognized as cells.

composed as a poem. On its face, *De rerum natura* ("On the nature of things") is contradictory: it appears to refute the very gods it also invokes. Perhaps the invocations, then, are "only" rhetorical, mere poetical curlicues. But I read the poem as suggesting something else:[11] if the gods exist, Lucretius's poem invites us to think, they are not made of atoms; they are not embodied; they are not capable of touching embodied things. Their eternality isolates them absolutely from the corporeal world. They thus function in the poem much as the void does, as that through which atoms forever fall. The possibilities of touching one another, of moving and being moved, of interacting, of communicating, of meaning – these are possibilities which require materiality, corporeality, finitude, death. Which is also to admit that they necessarily imply rhetoric. In not a fallen but a falling world,[12] rhetoric is the companion of motion, the unavoidable consequence of nothing remaining simply itself. So long as nothing stays the same, we can only speak and write and think rhetorically. This both opens up the vista of meaning, since what is not x can nonetheless stand for x; and it keeps that vista open, since what is not x can also stand for something else (as can x). Disclosure conceals as concealment discloses. Voila the tie between meaning's possibility and impossibility, the two as inseparable as faces of a coin. Meaning, like rainbows and horizons, can only be chased, not reached. Both its appearance and its

---

[11] Or, put otherwise, what follows is a sketch of what is at issue when something is "only" rhetorical.

[12] What an extraordinary accomplishment it would be – one moreover which has frequently been claimed with only the slightest fanfare – were our falling to be completed, perfected – were we simply to be fallen.

perpetual retreat are the consequence of our finitude, our inescapable situatedness.[13]

There is but one way to catch up with meaning. Only when the perpetual flight, or fall, of signification comes to an abrupt end is meaning's possibility absolutely and unequivocally achieved; but at that very moment its possibility is *indistinguishable* from its impossibility. Everything at last becomes equal to itself in being nothing.[14] Such an end is death, which both makes meaning possible and extinguishes it.[15] As gift, it is the promise of meaning – a promise denied the Lucretian gods, since what is deathless cannot mean. This is far from asserting, however, that what dies does mean. As with the perpetually plummeting Lucretian atoms, the pro-

---

[13] Put otherwise, what is meaning if not occasional, in the etymologic sense?

[14] I am alluding to Rousseau's completion of the circle of equality and inequality in part II of the Discourse on the Origin of Inequality. There he is describing the degeneration or, if you prefer, fall, of man in the state of society – but it is a motion that is at every moment driven by rhetorical possibility, rhetorical indeterminacy, which Rousseau confronts particularly in his notion of *perfectibilité*.

[15] One last time, one thinks of Hawthorne's "The birthmark." In his effort to rid his wife of the mark of her mortality, the mark which marked her as her birthmark, from the very beginning and before, Aylmer attained the perfection and the completeness which he sought at the cost it always demands: "The fatal hand had grappled with the mystery of life, and was the bond by which an angelic spirit kept itself in union with a mortal frame. As the last crimson tint of the birthmark – that sole token of human imperfection – faded from her cheek, the parting breath of the now perfect woman passed into the atmosphere, and her soul, lingering a moment near her husband, took its heavenward flight. Then a hoarse, chuckling laugh was heard again! Thus ever does the gross fatality of earth exult in its invariable triumph over the immortal essence which, in this dim sphere of half development, demands the completeness of a higher state."

mise of meaning requires constant falling, constant deferral. That deferral, as Darwin couldn't help but observe throughout On the Origin of Species, is paired to death, but it is either not completely ended by death – in which case meaning's plunge continues, however altered its trajectory – or it is – in which case meaning's plunge ends in utter annihilation.[16] To dream of a full presence and absolute subject which are not alienated from the possibility of meaning is to insist that man be man and god at once; it is to dream of an immortal god who also dies. That project may seem familiar, but its familiarity should not blind us to its intransigent, contradictory demand: what is required is an immortal, mortal god.

Superficially, the structure of this requirement parallels the pervasively unread autocriticism to which the essays of this book try to call attention: the god-like superhero who simultaneously does not merit serious scrutiny because he is, after all, only a comic book cliché; the master of knowledge reduced to cleaning a robot's diaper; the author whose ultimate responsibility has been atomized to didley-over-squat; the corrupting restorer or restoring corrupter; the essence which, become conditional or quasi, yet remains unabandoned. What are these if not chords vibrating in the mist, chords whose

---

16 Hence, on the Darwinian analogy, we have still-plunging lineages, represented either by living forms or by fossils. But we also have blanks, blanks which, however great our optimism, will never be filled; blanks so blank that we do not even know that they are there. One should also quickly add that the distinction between a plunge that continues and one that ends is always precarious and provisional; in a strange way, every plunge that continues is also a plunge that ends, and the affirmation of life – or rather, of *la survie* – is always also a work of mourning. The occasion of meaning, then, is not just a falling but a setting.

celestial high notes harmonize poorly with their terrestrial low notes? Far from attending to this harmony or disharmony, however, far from attending to the gaps separating notes always sounded together, those notes are instead individuated into strands of music unfolding over time. We hear the sequence of high notes and attend it as a melody which, we would believe, transcends the jarring and circumstantial cacophony over which it inessentially hovers. That melody emerges only by temporalizing the music, by understanding it sequentially, in terms of a past and a future. But this is to understand it rhetorically, with every note in some way standing for, calling back to or announcing another note, while simultaneously excusing our inattention to the concomitant cacophony. Temporality is thus to be understood as a mode of rhetoric, a mode of meaning's simultaneous promise and deferral. It is accordingly no accident at all that every example I have discussed in this book has in one form or another brushed against temporality as both problem and solution: whether in the form of the "fully-evolved" cell whose future is supposedly brighter than its monstrous cousins the parasite and the organelle; or in Darwin's diachronic understanding of the living beings whose good nature tends, though not into a distant futurity; or in the pursuit of evolutionary strategies which demand but cannot justify a determinate temporal horizon; or, finally, in the paired and seemingly neurotic endeavor to anticipate the future impact of scientific discoveries while simultaneously archiving what would be the entirety of their pasts. This resort to temporality, or rather rhetoric, to make sense of all of our projects is at once the assertion of a metaphysics which imagines temporality always in terms of presence – the past having once been present and the future being eventually present – and, in autocritical mode, the confession of the present absence of presence.

Driven to listen for a melody by the jarring gap between high and low notes, we no longer hear the interval, the provoking absence. It is precisely those intervals and absences, those gaps which we do not fail to experience as gaps, to which these essays, these fragments, these songs of mourning, are instead trying to call attention. They are essays in the old sense of "attempts"; they are fragments because what they attempt cannot be completed, as nothing can be completed so long as it falls; and they are sorrowful songs in that Orphic, lyrical mode which can only gesture ineffectually at the absences, whether as loss or lack, out of which it is born. Like Orpheus, these songs are ever turning back before they should. This last section, this last movement, is thus less conclusion than coda, whose impossible two-fold task is both to return the piece to its opening key and to provide a bridge from there into silence.

The first may seem less impossible than the second, but this is no small accomplishment of music, that an arbitrary key should over time assume the trappings of familiarity and home. "Ah! The tonic!" the sophisticated listener thinks. "We have happily ended at the beginning. It is time to clap now." We clap. We have not noticed the illusion which the musical piece has created, the illusion of a return, on the one hand, and of a completeness on the other. We have not noticed the silence which would be the last note, the note both of familiarity and uncanniness, relief and angst. Instead this silent note slips away unheard beneath the thundering applause. It is not clear what it would mean to listen for it. Nor, despite all pretense to the contrary, is there any preparation, musical or otherwise, for it. No final bridge transitions into silence seamlessly. Should we ever hear it, we will not know how to interpret it. Whether silence functions as a tonic or a dominant, whether it is resolution or screams for it – we will not know. Even the many efforts to instrumentalize silence in music – and what is music if not those attempts? – have failed and will always fail to master it. Silence breaks through; it interrupts; and its domestication in music is by temporalizing it, by promising its duration is foretold. But it isn't. It is always possible that a conductor will pause at a rest and never resume; just as it is always possible that a piece which appears to have ended has not. And silence, which interrupts, may itself be interrupted. That is in part why we are asked to turn off our cellphones at performances; but people still fidget in their seats, flip through the program, chatter idly with their companion; and even if they don't, in those moments which approach silence and stillness, the very sound of your neighbor's breathing is magnified, as is your own; and you hear perhaps the last faint and dying reverberation of a violin, unless you are remembering that rather than hearing it – but you can't tell. Silence is everywhere in music; but it is also nowhere. It is within the chord which would attempt to contain it between a high and a low note; it is between a first note and a second; it is marked as a rest or, with a double bar, as a conclusion. And yet, however pervasive its invocation, when is it ever found? When has a song truly been bracketed by silence, by a beginning out of nothing and an end into nothing? And equally, when has a song ever escaped silence? What would music be if there were no silence?

The visual correlate of silence is blankness, on which, in which, by which and through which we write. In a written text blankness appears as gaps, but in a rather special sense, for the gap was not created by the letters but rather was the condition of their possibility. As with silence in music, there is something threatening about the blank page – not least that in fact it fails to be contained by the page. Yet the blank page is also welcoming, and it may inspire relief as well as anxiety in us. But is it the blank page which inspires? Are we any more able to confront that blankness than we can confidently hear and interpret silence? Doesn't blankness loom as just as much an interruption – a break, a chasm – as silence? And as with silence does it not run the risk of itself being fractured, obliterated beneath the letter or the spilling ink or as blotting paper?

Why am I writing of silence and of blankness? In part to remind us that we live at their cusps; that the being that is ours, the almost-being that is ours, the being-not-quite that is ours, is poised precariously on this edge or the other of silence, on this edge or the other of blankness. "This edge or the other" because we cannot even say definitively on which side we find ourselves; we cannot say where the borders are or on what testimony we trust there are borders in the first place. This equivocation which as it were defines us is the equivocation which the metaphysics of presence summarily decides – and forgets. From within this forgetting there is never utter blankness or utter silence even in possibility; from within it we are never directionless. Thus, for example, progress is never put into question in any of the projects which I have critiqued. Who can doubt progress? Every project, every projection, is in the name of progress – even when each comes full circle. Thus progress now means killing hundreds of millions of maligned animals who threaten our notion of biodiversity, though a few hundred years ago progress meant deliberately introducing the same animals in order to advance civilization. Progress now means using science to deliver us from the energy crisis that science, in the name of progress, has in large part delivered us to.

On the cusp of silence or blankness, what we encounter instead is direction. That direction is always the accomplishment of rhetoric (*incognito*) whose own possibility depends on the silence and blankness that the accomplishment begins by denying. What would it mean to instead endure the possibility of silence and blankness? What would it mean to endure the absence of direction, not as exculpating but as demanding instead the highest responsibility? Such a thinking would be a thinking from within mourning, a thinking which no longer seeks to evade the absolute loss that is the condition of its possibility and which therefore loses the alibi of attempted preservation. Such a thinking acknowledges that preservation is not for us.[17]

I do not know if such a thinking is thinking at all. If it is, or even if it is not thinking but is something else, something on the far side of human possibility if it is anywhere at all, perhaps we find its example or its foreshadowing or at least

[17] And yet it does so not by embracing nihilism but by refuting it, by confronting nihilism with its secretly preservative project, its preservation of nothing as a positive fact.

something like its promise in Hamlet. For, like Hamlet, we are confronted by ghosts, by specters, by *what may not be there*. How shall we respond? What does it mean to respond? Is response possible? These are questions with which Hamlet wrestles. Hamlet wonders: Is the ghost real? Hamlet wonders: What is a real ghost? Hamlet wonders: What is a ghost of? Everything Hamlet does is in the context of *uncertain* answers to these questions. Being compelled and compelling himself[18] to act within that uncertainty in turn compels Hamlet to contemplate acting in terms of its equivocation, appropriately recognized by a gravedigger, between doing and performing; it drives him to recognize that this equivocation cannot absolutely and cannot ultimately be resolved; it forces him into the suspicion, which he can never put to rest whether by confirmation or refutation, that his is a part in a play.[19] Far from absolving Hamlet of responsibility, that suspicion exacerbates his responsibility: he rightly understands that he must and can only assume a responsibility for a script which he does not comprehend but which comprehends him.

Hamlet is not the only character in the play who confronts something like a ghost. But he is the only one – with the possible exception of Ophelia – who confronts it from the cusp of silence and blankness or, put otherwise, from within mourning. His foil, and ours, is Laertes. Students commonly

---

[18] Best would be a middle voice, a compulsion of which Hamlet is at once subject and object and not quite (or simply) either.

[19] Further, to borrow from a different Shakespearian play (*MacBeth*, Act V, Scene v) it forces him to confront himself as a poor player: "Life's but a walking shadow, a poor player / That struts and frets his hour upon the stage / And then is heard no more: it is a tale / Told by an idiot, full of sound and fury, signifying nothing."

prefer Laertes to Hamlet because Laertes is always *doing* something; Hamlet dithers, Hamlet – they say – never acts. But Laertes rushes home from France to avenge his murdered father and very nearly brings the kingdom of Denmark to its knees! He never doubts his duty nor his seeming knowledge, and his next step is always clear. Laertes, unlike Hamlet, never wants direction. Yet, in his impetuous action Laertes is always only following a script, one moreover which he falsely believes he comprehends. The responsibility he accepts is alibied by that mistaken comprehension and hence, ultimately, is no responsibility at all. He can claim, as he does, that the king is to blame. Laertes, in acting, has never confronted blankness or silence, has never understood himself as directionless and has never, thus, been responsible for the direction he has taken.

We too confront a blank and silent world of what might be ghosts – whether of a gigantic tortoise or of a rat, of a Pleistocene elephant or of our own unborn descendant. For the most part we dash busily about the stage, never lacking direction and contriving fencing matches with swords unbated and envenomed. In so doing we can thrive on the illusion of our agency and our mastery while simultaneously, like scientific authors, being absolved of true responsibility. Very well, then – let us be Laertes (or, for that matter, the Heideggerian "They") and commit atrocities we but dimly perceive and perhaps, in the last act, there will be someone else to blame. The alternative is aporia, a directionless anxiety where we grapple with questions we cannot answer and become responsible for what exceeds us. In short, the alternative is Hamlet's, full of angst and suffering and seemingly recommended only by the small circumstance that it is, perhaps, in some way truer. That may not be recommendation enough – and perhaps (again) ought not to be. Certainly it is far from clear that what is truer is also better, as it is far from clear that it matters at all whether something is truer or better. Hamlet, we might recall, does not evade absolute loss any more than Laertes.

That Hamlet and Laertes annihilate one another, like the mutual obliteration of matter and anti-matter, might invite us to question whether it is in fact truth which is at issue, whether it is truth, in fact, which, as small recommendation, distinguishes Hamlet's course. For surely we must wonder: Can we have a rigorous understanding of truth in a falling world, a world where meaning's possibility is inseparably conjoined to its impossibility, or does truth always smuggle with it a secret invocation of a transcendental signified and of fully achieved presence? The possibility which Hamlet marks is precisely only that – a possibility. It is not enduring, as truth is supposed to be, nor universal, nor certain. Instead it is circumstantial and terribly fragile. *The possibility which Hamlet marks is only a possibility and never more than a possibility.* Can Hamlet's alternative really be truer, then, or must we reconcile ourselves to an achievement far more tenuous and unsatisfying? We might say, perhaps too cavalierly, that Laertes, in ultimately not being responsible, ultimately did not respond, whether to his father or his father's ghost or his sister or his sister's ghost. With Hamlet the risks of so cavalier a verdict are the more pronounced; and that would be his success, to have accentuated and magnified those risks, such that we are the more uncertain how to answer the question – did he respond, is there in Hamlet a response to another, for another, from another? Did Hamlet preserve as possibility a possibility – and of what?

Is it enough that for Hamlet these are questions, questions which, no less than the ghost of a gigantic tortoise which we have but seemed to see, might succeed in demanding from us a response, if only a wail, a cry, a mournful dirge, the song of a rat or a goat or a swan, or an as-yet unbroken silence?[20]

Or – if this is different – is it for us to fall into the weeping brook, chanting snatches of old tunes as if incapable of our own distress, till our garments, heavy with their drink, pull us from our melodious lay to muddy death?

---

[20] I think of Czeslaw Milosz's poem "Meaning," particularly its last stanza, after it has been proposed that a thrush on a branch is just a thrush on a branch, not a sign:

> *–Even if that is so, there will remain*
> *A word wakened by lips that perish,*
> *A tireless messenger who runs and runs*
> *Through interstellar fields, through the revolving galaxies,*
> *And calls out, protests, screams.*

References

Abbott A (2000) German fraud inquiry casts a wider net of suspicion. *Nature* 405: 871 – 2.

Abbott A and Schwarz J (2002) Dubious data remain in print two years after misconduct inquiry. *Nature* 418: 113.

Abbott A, Cyranoski D, Jones N, Maher B, Schiermeier Q and van Noorden R (2010) Do metrics matter? *Nature* 465: 860 – 2.

Abdelkrim J, Pascal M and Samadi S (2005) Island colonization and founder effects: the invasion of the Guadeloupe islands by ship rats (*Rattus rattus*). *Molecular Ecology* 14: 2923 – 31.

Abedrabbo S (1994) Control of the Little Fire Ant, *Wasmannia auropunctata*, on Santa Fe island in the Galápagos islands. Pages 219 – 27 in Williams D (ed) Exotic Ants: Biology, Impact and Control of Introduced Species. Westview Press, Boulder.

Adam D and Knight J (2002) Publish and be damned. *Nature* 419: 772 – 6.

Akerley B, Rubin E, Novick V, Amaya K, Judson N and Mekalanos J (2002) A genome-scale analysis for identification of genes required for growth or survival of *Haemophilus influenzae*. *Proc Natl Acad Sci USA* 99: 966 – 971.

Akst J (2012) Anesthesiologist fabricates 172 papers. *The Scientist Magazine*, July 3.

Alpers S (1983) The Art of Describing. University of Chicago Press, Chicago.

Amann G, Stetter KO, Llobet-Brossa E, Amann R and Antón J (2000) Direct proof for the presence and expression of two 5% different 16S rRNA genes in individual cells of *Haloarcula marismortui*. *Extremophiles* 4: 373 – 376.

Anderson CJ and 43 others (2016) Response to Comment on estimating the reproducibility of psychological science. *Science* 351: 1037-c.

Anonymous (2013) Unreliable research: Trouble at the lab. *The Economist*, October 19.

Anonymous (2015a) Interdisciplinarity. *Nature* 525: 305

Anonymous (2015b) Credit and coauthors cause chatter. *Nature* 528: 11.

Archer M (2013) How we'll resurrect the gastric brooding frog, the Tasmanian tiger. TedTalk, March: http://www.ted.com/talks/michael_archer_how_we_ll_resurrect_the_gastric_brooding_frog_the_tasmanian_tiger?language=en.

Aristotle (2001) Metaphysics, Ross WD (trans). In McKeon R (ed) The Basic Works of Aristotle. The Modern Library, NY.

Aristotle (2001) The Physics, Hardie RP and Gaye RK (trans). In McKeon R (ed), *ibid.*

Associated Press (2012) Galápagos Islands steps up extermination of 180 million rats. *The Telegraph* (UK), November 15.

Australian Associated Press (2013) Extinct frog resurrected with 'de-extinction' technology. *The Guardian*, 21 November.

Bacon F (1960) The New Organon and Related Writings, Spedding J, Ellis R and Heath D (trans). The Liberal Arts Press, New York.

Baker M and Penny D (2016) Is there a reproducibility crisis? *Nature* 533: 452 – 4.

Bammer G (2015) Interdisciplinarity: Less vague please. *Nature* 526: 503.

Barber B (1961) Resistance by scientists to scientific discovery. *Science* 134: 596 – 602.

Baross JA and Hoffman SE (1985) Submarine hydrothermal vents and associated gradient environments as sites for the origin and evolution of life. *Origins of Life* 15: 327 – 45

Beasley MR, Datta S, Kogelnik H, Kroemer H and Monroe D (2002) Report of the investigation committee on the possibility of scientific misconduct in the work of Hendrik Schön and coauthors. Lucent Technologies, Murray Hill, New Jersey.

Begley CG and Ellis LM (2012) Raise standards for preclinical cancer research. *Nature* 483: 531 – 3.

Bell AW, Deutsch EW, Au CE, Kearney RE, Beavis R, Sechi S, Nilsson T, Bergeron JJM and HUPO Test Sample Working Group (2009) A HUPO test sample study reveals common problems in mass spectrometry-based proteomics. *Nature Methods* 6: 423 – 30.

Bellows K (2006) Galápagos Islands: Where the family of man meets nature. *National Geographic Traveler* April: 112 – 5.

Belshaw R, Pereira V, Katzourakis A, Talbot G, Pačes J, Burt A and Tristem M (2004) Long-term reinfection of the human genome by endogenous retroviruses. *Proc Natl Acad Sci USA* 101: 4894 – 9.

Belshaw R, Katzourakis A, Pačes J, Burt A and Tristem M. (2005) High copy number in human endogenous retrovirus families is associated with copying mechanisms in addition to reinfection. *Mol Biol Evol* 22: 814 – 7.

Benchley P (1999) Paradise in peril. *National Geographic* April: 2 – 31.

Bergson, H. (1998) Creative Evolution. Mitchell A (trans). Dover Publications Inc., Mineola, NY.

Blockstein DE (1998) Lyme disease and the passenger pigeon? *Science* 279: 1831.

Boersma P, Vargas H and Merlen G (2005) Living laboratory in peril. *Science* 308: 925.

Bohannon J (2013) Who's afraid of peer review? *Science* 342: 60 – 5.

Borges JL (1999) John Wilkins' Analytical Language. Pages 229 – 32 in Weinberger E (ed) Selected Non-fictions, Penguin Books, NY.

Borrell B (2009) A medical Madoff: Anesthesiologist faked data in 21 studies. *Scientific American*, March.

Bowman RI (1960) Report on a biological reconnaissance of the Galápagos Islands during 1957. United Nations Education, Scientific and Cultural Organization, Paris.

Boxma B, de Graaf R, van der Staay G, van Alen T and 11 others (2005) An anaerobic mitochondrion that produces hydrogen. *Nature* 434: 74 – 9.

Brosset A (1963) Statut actuel des mammifères des îles Galapagos. *Mammalia* 27: 323 – 40.

Brown R, Deletic A and Wong T (2015) How to catalyse collaboration. *Nature* 525: 315 – 7.

Brumfiel G (2002a) Time to wise up? *Nature* 418: 120 – 1.

Brumfiel G (2002b) Misconduct finding at Bell Labs shakes physics community. *Nature* 419: 419 – 21.

Brumfiel G (2002c) Physics guidelines drop equal-responsibility clause. *Nature* 420: 258.

Bull JJ and HA Wichman (2001) Applied evolution. *Annu Rev Ecol Syst* 32: 183 – 217.

Bursztyn M and Purushothaman S (2015) Inter-disciplinarity: Topping the charts. *Nature* 526: 323.

Button KS, Ioannidis JPA, Mokrysz C, Nosek BA, Flint J, Robinson ESJ and Munafò M (2013) Power failure: Why small sample size undermines the reliability of neuroscience. *Nature Reviews Neuroscience* 14: 365 – 76.

Cairns R (2011) A Critical Analysis of the Discourses of Conservation and Science on the Galápagos Islands. PhD dissertation, University of Leeds.

Calderón Alvarez C, Causton C, Hoddle M, Hoddle C, van Driesche R and Dtanek E (2012) Monitoring the effects of *Rodolia cardinalis* on *Icerya purchase* populations on the Galápagos Islands. *BioControl* 57: 167 – 79.

Callaham ML, Wears RL, Weber EJ, Barton C and Young G (1998) Positive outcome bias and other limitations in the outcome of research abstracts submitted to a scientific meeting. *JAMA* 280: 254 – 7.

Campbell K, Donlan CJ, Cruz F and Carrion V (2004) Eradication of feral goats *Capra hircus* from Pinta Island, Galápagos, Ecuador. *Oryx* 38: 328 – 33.

Caro T (2007) The Pleistocene re-wilding gambit. *TRENDS in Ecology and Evolution* 22: 281 – 3.

Carrion V, Donlan CJ, Campbell K, Lavoie C and Cruz F (2007) Feral donkey (*Equus asinus*) eradications in the Galápagos. *Biodiversity and Conservation* 16: 437 – 45.

Carrion V, Donlan CJ, Campbell K, Lavoie C and Cruz F (2011) Archipelago-wide island restoration in the Galápagos Islands: Reducing costs of invasive mammal eradication programs and reinvasion risk. *PLoS ONE* 6: e18835. doi: 10.1371/journal.pone.0018835.

Carroll SP, Jørgensen PS, Kinnison MT, Bergstrom CT, Denison RF, Gluckman P, Smith TB, Strauss SY, Tabashnik BE (2014) Applying evolutionary biology to address global challenges. *Science* 346: 1245993.

Causton C, Sevilla C and Porter S (2005) Eradication of the Little Fire Ant, *Wasmannia auropunctata* (Hymenoptera: Formicidae), from Marchena Island, Galápagos: On the edge of success? *Florida Entomologist* 88: 159 – 68.

CDF and WWF (2002) A biodiversity vision for the Galápagos Islands. Bensted-Smith R (ed). Charles Darwin Foundation, Puerto Ayora, Galápagos, Ecuador.

Check E (2002) Sitting in judgement. *Nature* 419: 332 – 3.

Clark DA and Clark DB (1981) Effects of seed dispersal by animals on the regeneration of *Bursera graveolens* (Burseraceae) on Santa Fe Island, Galápagos. *Oecologia* (Berl) 49: 73 – 5.

Clark DB (1980) Population ecology of *Rattus* across a desert-montane forest gradient in the Galápagos Islands. *Ecology* 61: 1422 – 33.

Clarke B, Murray J and Johnson M (1984) The extinction of endemic species by a program of biological control. *Pacific Science* 38: 97 – 104.

Clemens D and Johnson P (2000) Failure to detect DNA in hydrogenosomes of *Trichomonas vaginalis* by nick translation and immunomicroscopy. *Mol Biochem Parasitol* 106: 307 – 13.

Coetzee JM (2003) Elizabeth Costello. Viking, New York.

Cole S, Eiglmeier K, Parkhill J, James K and 40 others (2001) Massive gene decay in the leprosy bacillus. *Nature* 409: 1007 – 11.

Collins FS and Tabak LA (2014) NIH plans to enhance reproducibility. *Nature* 505: 612 – 3.

Cook J and Barrow J (1941) Captain Cook's Voyages of Discovery. J. M. Dent and Sons Ltd, London.

Corliss J, Dymond J, Gordon L, Edmond J, von Herzen R, Ballard R, Green K, Williams D, Bainbridge A, Crane K and van Andel T (1979) Submarine thermal springs on the Galápagos rift. *Science* 203: 1073 – 83.

Couzin-Frankel J (2013) The power of negative thinking. *Science* 342: 68 – 9.

Crick F (1954) The structure of the hereditary material. *Scientific American* 191: 54 – 61.

Crocker J and Cooper ML (2011) Addressing scientific fraud. *Science* 334: 1182.

Cruz F, Donlan CJ, Campbell K and Carrion V (2005) Conservation action in the Galápagos: Feral pig (*Sus scrofa*) eradication from Santiago Island. *Biological Conservation* 121: 473 – 8.

Cyranoski D (2004) Korea's stem-cell stars dogged by suspicion of ethical breach. *Nature* 429: 3.

Dalton R (2002) California lab fires physicist over retracted findings. *Nature* 418: 261.

Daruka I (2014) Confined to grow? Publication dynamics and the proliferation of scientific journals. *Europhysics News* 45: 19 – 22.

Darwin C (1962) The Voyage of the Beagle. Doubleday and Company, Garden City, NY.

Darwin C (2003) On the Origin of Species. A Facsimile of the First Edition. Harvard University Press, Cambridge.

De Groot R (1983) Tourism and conservation in the Galápagos Islands. *Biological Conservation* 26: 291 – 300.

Deleuze G (1998) Essays Critical and Clinical, Smith DW and Greco MA (trans). Verso, New York.

D'Elia M, Pereira M and Brown E (2009) Are essential genes really essential? *Trends Microbiol* 17 (10): 433 – 8.

Derrida J (1997) Of Grammatology (corrected edition), Spivak G (trans), Johns Hopkins University Press, Baltimore.

Descartes R (2006) Meditations Objections and Replies, Ariew R and Cress D (trans). Hackett Publishing, Indianapolis.

Dinerstein E and Irvin WR (2005) Rewilding: No need for exotics as natives return. *Nature* 437: 476.

Dirzo R, Young H, Galetti M, Ceballos G, Isaac N and Collen B (2014) Defaunation in the Anthropocene. *Science* 345: 401 – 6.

Donlan CJ and Martin P (2004) Role of ecological history in invasive species management and conservation. *Conservation Biology* 18: 267 – 9.

Donlan CJ, Greene H, Berger J, Bock C, Bock J, Burney D, Estes J, Foreman D, Martin P, Roemer G, Smith F, and Soulé M (2005) Re-wilding North America. *Nature* 436: 913 – 4.

Donlan CJ, Berger J, Bock C, Bock J, Burney D, Estes J, Foreman D, Martin P, Roemer G, Smith F, Soulé M and Greene H (2006) Pleistocene rewilding: An optimistic agenda for twenty-first century conservation. *The American Naturalist* 168: 660 – 81.

Donlan CJ, Campbell K, Cabrera W, Lavoie C, Carrion V and Cruz F (2007) Recovery of the Galápagos rail (*Laterallus spilonotus*) following the removal of invasive mammals. *Biological Conservation* 138: 520 – 4.

Dowler R, Carroll D and Edwards C (2000) Rediscovery of rodents (Genus *Nesoryzomys*) considered extinct in the Galápagos Islands. *Oryx* 34: 109 – 17.

Dufresne A, Salanoubat M, Partensky F, Artguenave F and 17 others (2003) Genome sequence of the cyanobacterium *Prochlorococcus marinus* SS120, a nearly minimal oxyphototrophic genome. *Proc Natl Acad Sci USA* 100: 10020 – 25.

Dyall S, Brown M and Johnson P (2004) Ancient invasions: From endosymbionts to organelles. *Science* 304: 253 – 7.

Eckhardt R (1972) Introduced plants and animals in the Galápagos Islands. *BioScience* 22: 585 – 590.

Editors (2002) Reflections on scientific fraud. *Nature* 419: 417.

Editors (2003) Retractions' realities. *Nature* 422: 1.

Editors (2013) The maze of impact metrics. *Nature* 502: 271.

Editors (2014) Number crunch. *Nature* 506: 131 – 2.

Editors (2015) Mind meld. *Nature* 525: 289.

Editors (2016) Reality check on reproducibility. *Nature* 533: 437.

Ehrlich P (2001) Intervening in evolution: Ethics and action. *Proc Natl Acad Sci USA* 98: 5477 – 5480.

Elias P (2005) Researcher tries to distance himself from cloning work. *The Seattle Times*, December 14.

Embley TM and Martin W (2006) Eukaryotic evolution, changes and challenges. *Nature* 440: 623 – 30.

ESPN films (2015) The case for de-extinction. Collector Series, November 25. http:// fivethirtyeight.com/features /the-scientist-trying-to-reverse-extinctions-think-jurassic-park/.

Evans JA (2013) Future science. *Science* 342: 44 – 5.

Evans J and Rzhetsky A (2010) Machine science. *Science* 329: 399 – 400.

Fanelli D (2009) How many scientists fabricate and falsify research? A systematic review and meta-analysis of survey data. *PLoS One* 4: e5738. doi: 10.1371/journal.pone. 0005738.

Fang FC, Steen RG and Casadevall A (2012) Misconduct accounts for the majority of retracted scientific publications. *Proc Natl Acad Sci USA* 109: 17028 – 33.

Fang G, Rocha E and Danchin A (2005) How essential are nonessential genes? *Mol Biol Evol* 22: 2147 – 56.

Ferguson C, Marcus A and Oransky I (2014) The peer-review scam. *Nature* 515: 480 – 2.

Fisher R (1955) Statistical methods and scientific induction. *Journal of the Royal Statistical Society Series B (Methodological)* 17: 69 – 78.

Flett I and Haberle S (2008) East of Easter: Traces of human impact in the far-eastern Pacific. Pages 281–299 in Clark G,

O'Connor S and Leach B (eds), Islands of Inquiry: Colonisation, Seafaring and the Archaeology of Maritime Landscapes. ANU E. Press, Canberra.

Folch J, Cocero MJ, Chesné P, Alabart JL, Domínguez V, Cognié Y, Roche A, Fernández-Arias A, Martí JI, Sánchez P and Echegoyen E (2009) First birth of an animal from an extinct species (*Capra pyrenaica pyrenaica*) by cloning. *Theriogenology* 71: 1026 – 34.

Friedmann EI and Ocampo R (1976) Endolithic blue-green algae in the Dry Valleys: Primary producers in the Antarctic desert ecosystem. *Science* 193: 1247 – 9.

Fry H (1979) Alexander Dalrymple and Captain Cook: The creative interplay of two careers. Pages 41–57 in Fisher R and Johnston H (eds) Captain James Cook and his Times. University of Washington Press, Seattle.

Gardels N (2015) Genome pioneer: We have the dangerous power to control evolution. An interview with genome and synthetic life scientist J. Craig Venter. *New Perspectives Quarterly* 32: 17 – 23.

Gerdes S, Scholle M, Campbell J, Balázsi G and 17 others (2003) Experimental determination and system level analysis of essential genes in *Escherichia coli* MG1655. *J Bacteriol* 185: 5673 – 84.

Gilbert DT, King G, Pettigrew S and Wilson TD (2016) Comment on Estimating the reproducibility of psychological science." *Science* 351: 1037-b.

Goetz T (2012) Life Hacker. *Wired* 20: 108 – 15.

Goodman SN (2016) Aligning statistical and scientific reasoning. *Science* 352: 1180 – 1.

Grant P (2002) Is a bell tolling for Bell Labs? *Nature* 417: 789.

Grant P and de Vries T (1993) The unnatural colonization of Galápagos by smooth-billed anis (*Crotophaga ani*). *Noticias de Galápagos* May: 21 – 3.

Greene M (2007) The demise of the lone author. *Science* 450: 1165.

Gunn W (2014) Fraud is not the big problem. *Nature* 505: 483.

Guo J (2006) The Galápagos Islands kiss their goat problem goodbye. *Science* 313: 1567.

Hadfield M and Mountain B (1981) A field study of a vanishing species, *Achatinella mustelina* (Gastropoda, Pulmonata), in the Waianae Mountains of Oahu. *Pacific Science* 34: 345 – 58.

Hamann O (1993) On vegetation recovery, goats and giant tortoises on Pinta Island, Galápagos, Ecuador. *Biodiversity and Conservation* 2: 138 – 51.

Hannaert V, Saavedra E, Duffieux F, Szikora JP, Rigden D, Michels P and Opperdoes F (2003) Plant-like traits associated with metabolism of *Trypanosoma* parasites. *Proc Natl Acad Sci USA* 100: 1067 – 71.

Harley D (2013) Scholarly communication: Cultural contexts, evolving models. *Science* 342: 80 – 2.

Harper G and Cabrera L (2010) Response of mice (*Mus musculus*) to the removal of black rats (*Rattus rattus*) in arid forest on Santa Cruz Island, Galápagos. *Biological Invasions* 12: 1449 – 52.

Harper G and Carrion V (2011) Introduced rodents in the Galápagos: Colonisation, removal and the future. Pages 63–66 in Veitch C, Clout M and Towns D (eds), Island Invasives: Eradication and Management. IUCN, Gland, Switzerland.

Harper G, Zabala J and Carrion V (2011) Monitoring of a population of Galápagos land iguanas (*Conolophus subcristatus*) during a rat eradication using brodifacoum. Pages 309–312 in Veitch, C., M. Clout and D. Towns, eds., Island Invasives: Eradication and Management. IUCN, Gland, Switzerland.

Harris D and Macdonald D (2007) Competition between introduced black rats and endemic Galápagos rice rats. *Ecology* 88: 2330 – 44.

Hawthorne N (2011) The birthmark. In Arvin N (ed), Hawthorne's Short Stories. Vintage Classics, NY.

Heidegger M (1962) Being and Time, Macquarrie J and Robinson E (trans). Harper Perennial, New York.

Heidegger M (1977) The Question Concerning Technology and Other Essays, Lovitt W (trans). Harper Perennial, New York.

Heidegger M. (1993) Basic Writings, Krell DF (ed). Harper Perennial, New York.

Hegel GWF (1956) The Philosophy of History, Sibree J (trans). Dover Publications, Mineola, NY.

Heinemann J (1999) Genetic evidence of protein transfer during bacterial conjugation. *Plasmid* 41: 240 – 7.

Hennessy E and McCleary A (2011) Nature's Eden? The production and effects of 'pristine' nature in the Galápagos Islands. *Island Studies Journal* 6: 131 – 56.

Hewitt A (2015) Man-made evolution is happening, and it's time to control it. http://newsroom.ucla.edu/releases/man-made-evolution-is-happening-and-its-time-to-control-it.

Hobbs R, Arico S, Aronson J, Baron J, Bridgewater P, Cramer V, Epstein P, Ewel J, Klink C, Lugo A, Norton D, Ojima D, Richardson D, Sanderson E, Valladares F, Vilà M, Zamora R and Zobel M (2006) Novel ecosystems: Theoretical and management aspects of the new ecological world order. *Global Ecology and Biogeography* 15: 1 – 7.

Hoffman PF, Kaufman AJ, Halverson GP and Schrag DP (1998) A Neoproterozoic Snowball Earth. *Science* 281: 1342 – 6.

Hotz RL (2015) How many scientists does it take to write a paper? Apparently, thousands. *The Wall Street Journal*, August 10.

Hough R (1994) Captain James Cook. W. W. Norton and Company, NY.

Howald G, Donlan CJ, Galván J, Russell J, Parkes J, Samaniego A, Wang Y, Veitch D, Genovesi P, Pascal M, Saunders A and Tershy B (2007) Invasive rodent eradication on islands. *Conservation Biology* 21: 1258 – 68.

Hoyman M and McCall J (2013) Is there trouble in paradise? The perspectives of Galápagos community leaders on managing economic development and environmental conservation through ecotourism policies and the Special Law of 1998. *Journal of Ecotourism* 12: 33 – 48.

Hutchison C, Peterson S, Gill S, Cline R, White O, Fraser C, Smith H and Venter JC (1999) Global transposon mutagenesis and a minimal Mycoplasma genome. *Science* 286: 2165 – 9.

Hutchison CA, Chuang RY, Noskov VN, Assad-Garcia N and 19 others (2016a) Design and synthesis of a minimal microbial genome. *Science* 351: aad6253.

Hutchison CA, Chuang RY, Noskov VN, Assad-Garcia N and 19 others (2016b) Research article summary: Design and synthesis of a minimal bacterial genome. *Science* 351: 1414.

Huxley J (1966) Charles Darwin: Galápagos and after. Pages 3–10 in Bowman R (ed), The Galápagos: Proceedings of the Symposia of the Galápagos International Science Project. University of California Press, Berkeley.

Ioannidis JPA (2005) Why most published research findings are false. *PLoS Med* 2: e124.

Ioannidis J. P. A. and 15 others (2009) Repeatability of published microarray gene expression analyses. *Nature Genetics* 14 (2): 149 – 55.

James M (2010) Collecting evolution: The vindication of Charles Darwin by the 1905-1906 Galápagos expedition of the California Academy of Sciences. *Proceedings of the California Academy of Sciences* 61: 197 – 210.

James WH (1995) Fraud and hoaxes in science. *Nature* 377: 474.

Jasny BR, Chin G, Chong L and Vignieri S (2011) Again, and again, and again… *Science* 334: 1225.

Jonas H (1966) The Phenomenon of Life: Toward a Philosophical Biology. Northwestern University Press, Evanston, Illinois.

Jørgensen D (2013) Reintroduction and de-extinction. *BioScience* 63: 719 – 20.

Juhas M, Eberl L and Glass J (2011) Essence of life: Essential genes of minimal genomes. *Trends Cell Biol* 21: 562 – 568.

Kafka F (1971) Report to an Academy. In Glatzer NN (ed), Franz Kafka: The Complete Stories. Schocken Books, NY.

Kaiser J (2001) Galápagos takes aim at alien invaders. *Science* 293: 590–592.

Katinka M, Duprat S, Cornillot E, Méténier G and 13 others (2001) Genome sequence and gene compaction of the eukaryote parasite *Encephalitozoon cuniculi*. *Nature* 414: 450 – 3.

Key G, Fielding A, Goulding M, Holm R and Stevens-Woods B (1998) Ship rats *Rattus rattus* on the Shiant Islands, Hebrides, Scotland. *Journal of Zoology* 245: 228 – 33.

King RD, Rowland J, Oliver SG, Young M, Aubrey W, Byrne E, Liakata M, Markham M, Pir P, Soldatova LN, Sparkes A, Whelan KE and Clare A (2009) The automation of science. *Science* 324: 85 – 9.

Kobayashi K, Ehrlich S, Albertini A, Amati G and 95 others (2003) Essential *Bacillus subtilis* genes. *Proc Natl Acad Sci USA* 100: 4678 – 83.

Koonin E (2000) How many genes can make a cell: The minimal gene set concept. *Annu Rev Genom Human Genet* 1: 99 – 116.

Koonin E (2003) Comparative genomics, minimal gene-sets, and the last universal common ancestor. *Nature Rev Microbiol* 1: 127 – 36.

Kramer P (1983) The Galápagos: Islands under siege. *Ambio* 12: 186 – 90.

Kramer P (1984) Man and other introduced organisms. *Biological Journal of the Linnean Society* 21: 253 – 8.

Krell F and Wheeler Q (2014) Specimen collection: Plan for the future. *Science* 344: 815 – 6.

Kristof N (2005) Where deer and lions play. *New York Times*, December 13.

Lambdin C (2012) Significance tests as sorcery: Science is empirical – significance tests are not. *Theory Psychology* 22: 67 – 90.

Langmuir I (1989) Pathological science. (Transcript of a talk given at The Knolls Research Laboratory, Schenectady, New York, NY, December 18, 1953) *Physics Today* 42: 36 – 48.

Larsen PO and von Ins M (2010) The rate of growth in scientific publication and the decline in coverage provided by Science Citation Index. *Scientometrics* 84: 575 – 603.

Lavoie C, Cruz F, Carrion V, Campbell K, Donlan CJ, Harcourt S and Moya M (2007) The Thematic Atlas of Project Isabela. Charles Darwin Foundation, Puerto Ayora, Galápagos, Ecuador.

Lavoisier A (1965) Elements of Chemistry, Kerr R (trans). Dover Publications Inc., NY.

Lawrence PA (2002) Rank injustice. *Nature* 415: 835 – 6.

Lawton J (1997) The science and non-science of conservation biology. *Oikos* 79: 3 – 5.

Lazdunski C, Bouveret E, Rigal A, Journet L, Lloubès R and Bénédetti H (1998) Colicin import into *Escherichia coli* cells." *J Bacteriol* 180: 4994 – 5002.

Le Breton Y, Belew A, Valdes K, Islam E and 5 others (2015) Essential genes in the core genome of the human pathogen *Streptococcus pyogenes*. *Sci Rep* 5: 9838. doi: 10.1038/srep09838.

Lederberg J (1952) Cell genetics and hereditary endosymbiosis. *Physiological Reviews* 32: 403 – 30.

Ledford H (2015) Team Science. *Nature* 525: 308 – 11.

León-Avila G and Tovar J (2004) Mitosomes of *Entamoeba histolytica* are abundant mitochondrion-related remnant organelles that lack a detectable organellar genome. *Microbiology* 150: 1245 – 50.

Lipton P (2005) Testing hypotheses: Prediction and prejudice. *Science* 307: 219 – 21.

Loope L, Hamann O and Stone C (1988) Conservation biology of oceanic archipelagoes: Hawaii and the Galápagos. *BioScience* 38: 272 – 82.

Lucretius (1995) On the Nature of Things, Esolin AM (trans). Johns Hopkins University Press, Baltimore.

Luo H, Lin Y, Gao F, Zhang C and Zhang R (2014) DEG 10, an update of the database of essential genes that includes both protein-coding genes and noncoding genomic elements. *Nucl Acids Res* 42: D574 – 80.

Lwoff, A (1957) The concept of virus. *J Gen Microbiol* 17: 239 – 53.

Mabe M and Amin M (2001) Growth dynamics of scholarly and scientific journals. *Scientometrics* 51: 147 – 62.

Mahato RN, Lülf H, Siekman MH, Kersten SP, Bobbert PA, de Jong MP, De Cola L and van der Wiel WG (2015) Retraction. *Science* 350: 170.

Maimonides M. (1963) The Guide of the Perplexed, Pines S (trans). University of Chicago Press, Chicago.

Martin ED (1926) The Meaning of a Liberal Education. Norton and Co, NY.

Martinson BC, Anderson MS and de Vries R (2005) Scientists behaving badly. *Nature* 435: 737 – 8.

Maugh TH (2008) Physicist is found guilty of misconduct. *Los Angeles Times*, July 19.

McDonald RJ, Cloft HJ and Kallmes DF (2007) Fate of submitted manuscripts rejected from the *American Journal of Neuroradiology*: Outcomes and commentary. *AJNR* 28: 1430 – 4.

McNutt M (2013) Improving scientific communication. *Science* 342: 13.

Melville H (1969) Great Short Works of Herman Melville, Berthoff W (ed). Perennial Classics, NY.

Mills D, Peterson R and Spiegelman S (1967) An extracellular Darwinian experiment with a self-duplicating nucleic acid molecule. *Proc Natl Acad Sci USA* 58: 217 – 24.

Minteer B, Collins J, Love K and Puschendorf R (2014) Avoiding (re)extinction. *Science* 344: 260 – 1.

Moreira D and López-García P (2009) Ten reasons to exclude viruses from the tree of life. *Nature Rev Microbiol* 7: 306 – 11.

Morowitz HJ (1968) Energy Flow in Biology. Academic Press, NY.

Mushegian A (1999) The minimal genome concept. *Curr Opin Gen Devel* 9: 709 – 14.

Mushegian A and Koonin E (1996) A minimal gene set for cellular life derived by comparison of complete bacterial genomes. *Proc Natl Acad Sci USA* 93: 10268 – 73.

Myers N and Knoll A (2001) The biotic crisis and the future of evolution. *Proc Natl Acad Sci USA* 98: 5389 – 92.

Nash S (2008) Ecotourism and other invasions. *BioScience* 59: 106 – 10.

Nemery B (2001) What happens to the manuscripts that have not been accepted for publication in *Occupational and Environmental Medicine*? *Occup Environ Med* 58: 604 – 7.

Nicholls H (2013) The 18-km^2 rat trap. *Nature* 497: 306 – 8.

Nietzsche F (1988) Beyond Good and Evil. Faber M (trans), Oxford University Press, Oxford.

Nietzsche F (2012) On Truth and Lies in a Nonmoral Sense. Nikolic D (trans). Aristeus Books, Chicago.

Nosek BA, Alter G, Banks GC, Borsboom D and 35 others (2015) Promoting an open research culture. *Science* 348: 1422 – 5.

Nuzzo R (2014) Statistical errors. *Nature* 506: 150 – 2.

Oehlenschläger F and Eigen M (1997) 30 years later – A new approach to Sol Spiegelman's and Leslie Orgel's *in vitro* evolutionary studies. *Orig Life Evol Biosph* 27: 437 – 57.

Oliveira-Santos L and Fernandez F (2010) Pleistocene rewilding, Frankenstein ecosystems, and an alternative conservation agenda. *Conservation Biology* 24: 4 – 6.

Open Science Collaboration (2015) Estimating the reproducibility of psychological science. *Science* 349: aac4716.

Plato (1997) The Symposium, Nehamas A and Woodruff P (trans). In Cooper JM (ed), *ibid.*

Pormann PE (2015) Inside Manchester's 'arts lab'. *Nature* 525: 318 – 9.

Primack RB (2013) Scientific misconduct occurs, but is rare. *Biological Conservation* 157: iii.

Prinz F, Schlange T and Asadullah K (2011) Believe it or not: How much can we rely on published data on potential drug targets? *Nature Reviews Drug Discovery* 10: 712.

Rabesandratana T (2013) The seer of science publishing. *Science* 342: 66 – 7.

Ray J, Berkwits M and Davidoff F (2000) The fate of manuscripts rejected by a general medical journal. *Am J Med* 109: 131 – 5.

Reich ES (2013) The golden club. *Nature* 502: 291 – 3.

Rennie D. and Flanagin A (1994) Authorship! Authorship! Guests, ghosts, grafters and the two-sided coin. *JAMA* 271: 469 – 71.

Robbins B and Ross A (1996) Mystery science theater (with a reply by A. D. Sokal). *Lingua Franca* 6: 54 – 7.

Rocha LA, Aleixo A, Allen G, Almeda F and 119 co-authors (2014) Specimen collection: An essential tool. *Science* 344: 814 – 5.

Roemer G, Donlan CJ and Courchamp F (2002) Golden eagles, feral pigs, and insular carnivores: How exotic species turn native predators into prey. *Proc Natl Acad Sci USA* 99: 791 – 6.

Rosenberg D, Wilson M and Cruz F (1990) The distribution and abundance of the smooth-billed ani *Crotophaga ani* (L.) in the Galápagos Islands, Ecuador. *Biological Conservation* 51: 113 – 23.

Ross JS, Hill KP, Egilman DS and Krumholz HM (2008) Guest authorship and ghost writing in publications related to Rofecoxib: A case study of industry documents from Rofecoxib litigation. *JAMA* 299: 1800 – 12.

Rothman D, Fournier G, French K, Alm E, Boyle E, Cao C and Summons R (2014) Methanogenic burst in the end-Permian carbon cycle. *Proc Natl Acad Sci USA* 111: 5462 – 7.

Rubenstein DR, Rubenstein DI, Sherman P and Gavin T (2006) Pleistocene park: Does re-wilding North America represent sound conservation for the 21st century? *Biological Conservation* 132: 232 – 8.

Ryan MJ (2011) Replication in field biology: The case of the frog-eating bat. *Science* 334: 1229 – 30.

Rylance R (2015) Global funders to focus on interdisciplinarity. *Nature* 525: 313 – 5.

Rzhetsky A, Iossifov I, Loh JM and White KP (2006) Microparadigms: Chains of collective reasoning in publications about molecular interactions. *Proc Natl Acad Sci USA* 103: 4940 – 5.

Saey TH (2015) Is redoing scientific research the best way to find truth? *Science News* 187: 20.

Sassetti C, Boyd D and Rubin E (2001) Comprehensive identification of conditionally essential genes in Myco-bacteria. *Proc Natl Acad Sci USA* 98: 12712 – 7.

Saul H (2013) Extinct frog brought back to life. *The Independent (UK)*, 22 November.

Schlaepfer, M. (2005) Re-wilding: A bold plan that needs native megafauna. *Nature* 437: 951.

Schön JH, Kloc Ch and Batlogg B (2003) Retraction of Superconductivity in molecular crystals induced by charge injection. *Nature* 422: 93.

Schön JH, Kloc Ch, Bucher E and Batlogg B (2003) Retraction of Efficient organic photovoltaic diodes based on doped pentacene. *Nature* 422: 93.

Schramski J, Gattie D and Brown J (2015) Human domination of the biosphere: Rapid discharge of the earth-space battery foretells the future of humankind. *Proc Natl Acad Sci USA* 112: 9511 – 7.

Schroter S, Black N, Evans S, Godlee F, Osorio L and Smith R (2008) What errors do peer reviewers detect, and does training improve their ability to detect them? *J R Soc Med* 101: 507 – 14.

Seddon P, Griffiths C, Soorae P and Armstrong D (2014) Reversing defaunation: Restoring species in a changing world. *Science* 345: 406 – 12.

Service RF (2016) Synthetic microbe has fewest genes but many mysteries. *Science* 351: 1380 – 1.

Shay S (2005) Rewilding: Don't overlook humans living on the plains. *Nature* 437: 476.

Sherkow JS and Greely HT (2013) What if extinction is not forever? *Science* 340: 32 – 3.

Simberloff D, Genovesi P, Pyšek P and Campbell K (2011) Recognizing conservation success. *Science* 332: 419 – 21.

Simkin M and Roychowdhury V (2006) Do you sincerely want to be cited? Or: read before you cite. *Significance* 3: 179 – 81.

Simmons JP, Nelson LD and Simonsohn U (2011) False-positive psychology: Undisclosed flexibility in data collection and analysis allows presenting anything as significant. *Psychological Science* 22: 1359 – 66.

Smaglik P (2002) Learning from misconduct. *Nature* 420: 3.

Smith C (2005) Re-wilding: Introductions could reduce biodiversity. *Nature* 437: 318.

Sokal AD (1996) A physicist experiments with cultural studies. *Lingua Franca* 6: 62 – 4.

Spotts P (2015) An emerging challenge to science's credibility. *The Christian Science Monitor*, August 28.

Stahl M and Stintzi A (2015) Microarray transposon tracking for the mapping of conditionally essential genes in *Campylobacter jejuni*. *Methods Mol Biol* 1279: 1 – 14.

Stapel DA and Lindenberg S (2011) Retraction. *Science* 334: 1202.

Steen RG (2011) Retractions in the scientific literature: Is the incidence of research fraud increasing? *J Med Ethics* 37: 249 – 53.

Stokstad E (2015) Bring back the aurochs. *Science* 350: 1144 – 7.

Stone R and Jasny B (2013) Scientific discourse: Buckling at the seams. *Science* 342: 57.

Stott R (2003) Darwin and the Barnacle. W. W. Norton and Company, New York.

Szostak R. (2015) Interdisciplinarity: Resources abound. *Nature* 526: 503.

Tamas I, Klasson L, Canbäck B, Näslund AK and 5 others (2002) 50 million years of genomic stasis in endosymbiotic bacteria. *Science* 296: 2376 – 9.

Tewksbury J and Rogers H (2014) An animal-rich future. *Science* 345: 400.

Thornton I (1971) Darwin's Islands: A Natural History of the Galápagos. The Natural History Press, Garden City, NY.

Tovar J, León-Avila G, Sánchez L, Sutak R and 5 others (2003) Mitochondrial remnant organelles of *Giardia* function in iron-sulphur protein maturation. *Nature* 426: 172 – 6.

Twilley N (2016) What's the point of streamlining nature? *The New Yorker*: April 2.

van der Giezen M, Sjollema K, Artz R, Alkema W and Prins R (1997) Hydrogenosomes in the anaerobic fungus *Neocallimastix frontalis* have a double membrane but lack an associated organelle genome. *FEBS Lett* 408: 147 – 50.

van Noorden R (2010) A profusion of measures. *Nature* 465: 864 – 6.

Villarreal L and Witzany G (2010) Viruses are essential agents within the roots and stem of the tree of life. *J Theor Biol* 262: 698 – 710.

Vince G (2011) Embracing invasives. *Science* 331: 1383 – 4.

Viseu A (2015) Integration of social science into research is critical. *Nature* 525: 291.

Vogel G (2011) Psychologist accused of fraud on 'astonishing scale.' *Science* 334: 579.

von Uexküll J (2010) A Foray into the Worlds of Animals and Humans: With a Theory of Meaning, O'Neil JD (trans). University of Minnesota Press, Minneapolis.

Waltz D and Buchanan BG (2009) Automating science. *Science* 324: 43 – 4.

Wang D, Song C and Barabási AL (2013) Quantifying long-term scientific impact. *Science* 342: 127 – 32.

Wassen MJ and Heckert MP (2015) Interdisciplinarity: Two more principles. *Nature* 526: 323.

Waters E, Hohn M, Ahel I, Graham D and 18 others (2003) The genome of *Nanoarchaeum equitans*: Insights into early Archaeal evolution and derived parasitism. *Proc Natl Acad Sci USA* 100: 12984 – 8.

Wells LE (2006) Viral Adaptations to Life in the Cold. PhD dissertation, University of Washington, Seattle.

Western D (2001) Human-modified ecosystems and future evolution. *Proc Natl Acad Sci USA* 98: 5458 – 65.

Wetterer JK and Porter SD (2003) The little fire ant, *Wasmannia auropunctata*: Distribution, impact and control. *Sociobiology* 41: 1–41.

Wilkins B and Thomas A (2000) DNA-independent transport of plasmid primase protein between bacteria by the I1 conjugation system. *Mol Microbiol* 38: 650 – 7.

Williams B, Hirt R, Lucocq J and Embley TM (2002) A mitochondrial remnant in the microsporidian *Trachipleistophora hominis*. *Nature* 418: 865 – 9.

Woese CR (1998) The universal ancestor. *Proc Natl Acad Sci USA* 95: 6854 – 9.

Woese CR (2000) Interpreting the universal phylogenetic tree. *Proc Natl Acad Sci USA* 97: 8392 – 6.

Woese CR (2002) On the evolution of cells. *Proc Natl Acad Sci USA* 99: 8742 – 7.

Woese CR (2005) Evolving biological organization. Pages 99 – 117 in Sapp J (ed), Microbial Phylogeny and Evolution: Concepts and Controversies, Oxford University Press, Oxford.

Woese CR and Goldenfeld N (2009) How the microbial world saved evolution from the Scylla of molecular biology and the Charybdis of the modern synthesis. *Microbiol Mol Biol Rev* 73: 14 – 21.

Woolf V (1981) Mrs. Dalloway, Harcourt Brace Jovanovich Publishers, San Diego, New York and London.

Zavaleta E, Hobbs R and Mooney H (2001) Viewing invasive species removal in a whole-ecosystem context. *TRENDS in Ecology and Evolution* 16: 454 – 9.

Zimmer C (2013) Bringing them back to life. *National Geographic* 223: 28 – 41.

Made in the USA
Monee, IL
05 November 2026

67383823R00238